UMAP

Modules

Tools for Teaching 1987

published by
Consortium for Mathematics and
 Its Applications, Inc.
60 Lowell Street
Arlington, MA 02174

edited by
Paul J. Campbell

Beloit College
Campus Box 194
705 College Street
Beloit, WI 53511

All rights reserved. No part of this publication may be reproduced, stored in a retrieval system, or transmitted in any form or by any means, electronic, mechanical, photocopying, recording, or otherwise, without prior permission of the copyright holder.

This material was prepared with the partial support of National Science Foundation Grants No. SED80-07731 and SPE-8304192. Recommendations expressed are those of the authors and do not necessarily reflect the views of the NSF or the copyright holder.

©1988 COMAP, Inc. All rights reserved.
ISBN 0-912843-12-8
Printed in USA

Table of Contents

Introduction v

Unit		Page
318	Tomography: Three-Dimensional Image Reconstruction .... 1 Frederick Solomon	
474	Nominal vs. Effective Rates of Interest .... 21 Peter A. Lindstrom	
518	Oligopolistic Competition .... 55 Donald R. Sherbert	
526	Dimensional Analysis .... 71 Frank R. Giordano, Michael E. Wells, Carroll O. Wilde	
530	Perturbation Theory .... 99 Bertram Ross	
576	Randomized Response Technique: Getting in Touch With Touchy Questions .... 125 Paul Mullenix	
674	Price Elasticity of Demand .... 151 Yves Nievergelt	
675	The Lotka-Volterra Predator-Prey Model .... 183 James Morrow	
676	Compartment Models in Biology .... 207 Ron Barnes	
677	Funding Pension Benefits .... 235 Ho Kuen Ng	
679	The Solar Concentrating Properties of a Conical Reflector .... 251 Don Leake	
681	Simple Mortality Functions .... 279 Ho Kuen Ng	
684	Linear Programming Via Elementary Matrices .... 293 Helen Wang	

Introduction

The instructional modules in this volume were developed by the Undergraduate Mathematics and Its Applications (UMAP) Project. UMAP has been funded by grants from the National Science Foundation to Education Development Center, Inc. (1976–February 1983) and to the Consortium for Mathematics and Its Applications (COMAP), Inc. (February 1983–February 1985). Project UMAP develops and disseminates instructional modules and expository monographs in the mathematical sciences and their applications for undergraduate students and instructors.

UMAP modules are self-contained (except for stated prerequisites), lesson-length, instructional units from which undergraduate students learn professional applications of mathematics and statistics to such fields as biomathematics, economics, American politics, numerical methods, computer science, earth science, social sciences, and psychology. The modules are written and reviewed by classroom instructors in colleges and high schools throughout the United States and abroad. In addition, a number of people from industry are involved in the development of instructional modules.

In addition to the annual collection of UMAP modules, COMAP also distributes individual UMAP instructional modules, *The UMAP Journal*, and the UMAP expository monograph series. Thousands of instructors and students have shared their reactions to the use of these instructional methods in the classroom. Comments and suggestions for changes are incorporated as part of the development and improvement of materials.

The substance and momentum of the UMAP Project comes from the thousands of individuals involved in the development and use of UMAP's instructional materials. In order to capture this momentum and succeed beyond the period of federal funding, we established COMAP as a nonprofit organization. COMAP is committed to the improvement of mathematics education, to the continuation of the development and dissemination of instructional materials, and to fostering and enlarging the network of people involved in the development and use of materials. COMAP deals with science and mathematics education in secondary schools, teacher training, continuing education, and industrial and government training programs.

Incorporated in 1980, COMAP is governed by a Board of Trustees:

David Roselle, President	VA Polytechnic Institute and State University
Robert Thrall, Treasurer	Rice University
William Lucas, Clerk	Claremont Graduate School
Joseph Malkevitch	York College, CUNY
Henry Pollak	Bell Communications Research, Retired
Trudi C. Miller	University of Minnesota
Marion T. Jones	Devitt Jones Productions, Ltd.
H. Newton Garber	Merrill, Lynch and Co.

Instructional programs are guided by the Consortium Council, whose members are variously elected by the broad COMAP membership, or appointed by cooperating organizations (Mathematical Association of America, Society for Industrial and Applied Mathematics, National Council of Teachers of Mathematics, the American Mathematical Association of Two-Year Colleges, and The Institute of Management Sciences, and the American Statistical Association).The 1988 Consortium Council is chaired by Margaret Cozzens (Northeastern University), and its members are:

Margaret Barry Cozzens, Chair	Northeastern University
Robert Barnes	University of Houston
Alphonese Buccino	University of Georgia
Paul J. Campbell	Beloit College
Toni Carroll	Siena Heights College
Gary Froelich	Bismarck H.S., Bismarck, ND
Ben Fusaro	U.S. Military Academy
Frank R. Giordano	U.S. Military Academy
Landy Godbold	The Westminster Schools
JoAnne Growney	Bloomsburg University
Charles Hamberg	Illinois St. Math/Science Academy
Irwin Hoffman	Geo. Washington H.S., Denver
Zaven Karian	Denison University
Walter Meyer	Adelphi University
Peter Lindstrom	North Lake College
William Lucas	Claremont Graduate School
Warren Page	NYC Technical College, CUNY
Fred S. Roberts	Rutgers University
Stephen Rodi	Austin Community College
Robert T. Shanks	Edison Public Schools, Edison, NJ
Gene Woolsey	Colorado School of Mines

This collection of modules represents the spirit and ability of scores of volunteer authors, reviewers, and field-testers (both instructors and students). The modules also present various fields of application as well as different levels of mathematics. COMAP is very interested in receiving information on the use of modules in various settings. We invite you to contact us.

UMAP

Modules in Undergraduate Mathematics and its Applications

Published in cooperation with the Society for Industrial and Applied Mathematics, the Mathematical Association of America, the National Council of Teachers of Mathematics, the American Mathematical Association of Two-Year Colleges, The Institute of Management Sciences, and the American Statistical Association.

Module 318

Tomography: Three-Dimensional Image Reconstruction

Frederick Solomon

INTERMODULAR DESCRIPTION SHEET:	UMAP Unit 318
TITLE:	Tomography: Three Dimensional Image Reconstruction
AUTHOR:	Frederick Solomon Warren Wilson College Swannanoa, NC 28778
REVIEW STAGE / DATE:	III 1/23/79
CLASSIFICATION:	Appl Analysis/Medical Radiology
SUGGESTED SUPPORT MATERIAL:	
REFERENCES:	See Section 9 of text.
PREREQUISITE SKILLS:	1. Familiarity with the definition and basic properties of Fourier Transforms. 2. Familiarity with contour integration and the Residue Theorem in the complex plane.
OUTPUT SKILLS:	1. Acquaintance with a significant applied math problem utilizing Fourier Transforms. 2. Generalization of the Fourier Transforms to two dimensions. 3. Practice with Fourier Transforms. 4. Introduction to the Hankel Transform.

© 1979, 1988 by COMAP, Inc. All rights reserved.

COMAP, 60 Lowell Street, Arlington, MA 02174 (617) 641-2600

Tomography: Three-Dimensional Image Reconstruction

Frederick Solomon
Division of Natural Science
State University of New York College at Purchase
Purchase, NY 10577

Table of Contents

1. THE PROBLEM 1
2. NOTATION 2
3. BACK PROJECTION: AN APPROXIMATION 4
4. FOURIER TRANSFORM NOTATION 6
5. THE EXACT SOLUTION 7
6. THE BASIC THEOREM 8
7. CIRCULAR SYMMETRY 10
8. THE CIRCULARLY ASYMMETRIC CASE 13
9. HISTORICAL AND BIBLIOGRAPHICAL NOTE 14
10. ANSWERS AND SUGGESTIONS TO EXERCISES 14

Modules and Monographs in Undergraduate Mathematics and Its Applications (UMAP) Project

The goal of UMAP was to develop, through a community of users a developers, a system of instructional modules in undergraduate mathema and its applications to be used to supplement existing courses and fr which complete courses may eventually be built.

The Project was guided by a National Advisory Board of mathem cians, scientists, and educators. UMAP was funded by a grant from National Science Foundation and is now supported by the Consortium Mathematics and Its Applications (COMAP), Inc., a non-profit corpora engaged in research and development in mathematics education.

COMAP Staff

Paul J. Campbell	Editor
Solomon A. Garfunkel	Executive Director, COMAP
Laurie W. Aragon	Business Development Manager
Philip A. McGaw	Production Manager
Theresa Cronin	Copy Editor
Annemarie S. Morgan	Administrative Assistant
John Gately	Distribution

The Project would like to thank R. Bruce Mericle, William Glessner and Robert L. Baker, Jr. for the reviews, and all others who assisted in the production of this unit.

This material was prepared with the support of National Science Foundation Grant No. SE 76-19615 A02. Recommendations expressed are those of the author and do not necessarily reflect the vie of the NSF, nor of the National Steering Committee.

1. The Problem

Suppose we want to map out the interior of an opaque object. To be specific, suppose we want to determine the interior of part of a human body—say, an arm. An x-ray photograph reveals the two-dimensional projection of the arm on a viewing screen. This projection indicates points of high and low mass density. However, given a point of high density, we still do not know where in the interior of the arm the point is located—it could be anywhere on the line perpendicular to the screen and hence parallel to the x-ray beam. The problem is an important one. A lesion in the brain will register on the viewing screen; and it is important to know just where along the line parallel to the x-ray beam the lesion is located. So the problem is in constructing a three-dimensional map from two-dimensional pictures. (Actually, the phrase "mass density" is used in a suggestive sense here; it is only one factor affecting x-ray impenetrability.)

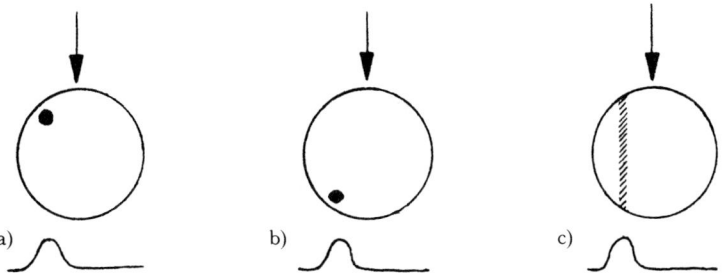

Figure 1. Arrows indicate direction of x-ray beam on a circular object. The dark area inside the object corresponds to a region of high mass density. All three have the same projection on the viewing screen, but the locations of the high mass density are quite different. It might even be uniformly spread along the line parallel to the x-ray beam, as shown in (c).

As we shall see, if x-ray photographs are taken at many different angles, then the information from all of them can be synthesized into the desired three-dimensional map. The purpose of this unit is to show how this is done.

"...the goal is to construct a three-dimensional map...."

Although the goal is to construct a three-dimensional map, the problem is actually a two-dimensional one. If x, y, and z-axes are situated so that the x-ray beams are perpendicular to the z-axis, it is enough to construct a map of each cross section parallel to the xy plane. Thus we will be concerned solely with a two-dimensional object in the xy plane with incident x-rays in this plane.

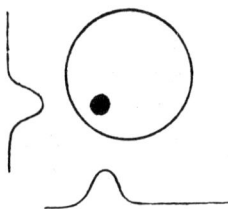

Figure 2. Two x-ray projections indicate the point of high mass density.

2. Notation

Let $f(x, y)$ be the mass density of the object and let $\phi + \pi/2$ be the angle between the x-axis and the incident beam of x-rays. Draw u, v-axes as in **Figure 3**, that is, the v-axis points in the

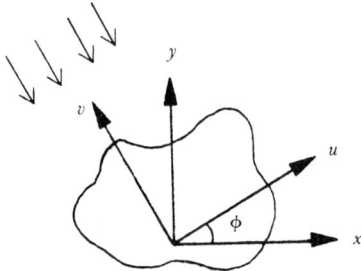

Figure 3. Arrows in upper left indicate the direction of x-rays.

direction of the x-ray source, while the u-axis is perpendicular to the v-axis, with the u, v-axes in the same orientation as the x, y-axes. Finally we define the *ray-sum* $R(u, \phi)$ as the *total* mass registered u units from the v-axis when $\phi + \pi/2$ is the angle of the x-ray beam. That is, $R(u_0, \phi_0)$ is the total mass on line l in **Figure 4(b)**.

Now we assume that all ray-sums $R(u, \phi)$ (for all u and all ϕ) are known. (In actuality we would know the function $R(u, \phi)$ for a finite number of angles ϕ, as we direct beams from different direc-

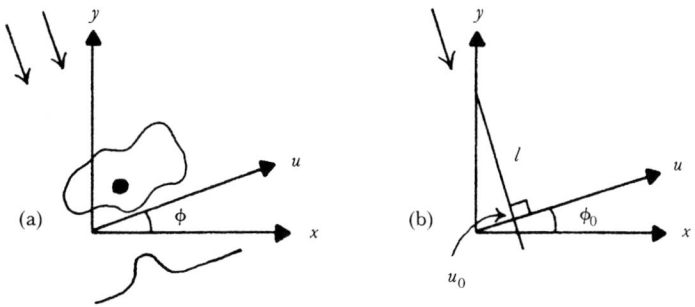

Figure 4(a). Graph of $R(u, \phi)$ for fixed ϕ. **(b)** Line l is u_0 from origin.

tions at the object; and $R(u, \phi)$ would be approximated by interpolation for all other values of ϕ.) Notice that $R(u, \phi + \pi) = R(-u, \phi)$ (Why?); so it is only necessary to find $R(u, \phi)$ for $0 \leq \phi \leq \pi$.

The definition for fixed u_0 and ϕ_0 of $R(u_0, \phi_0)$ is related to the mass density $f(x, y)$ by

$$R(u_0, \phi_0) = \text{total mass along } l \text{ in \textbf{Figure 4(b)}}$$

$$= \int_l f(x, y) \, ds,$$

where s is arclength along l. The equation of l is, of course, just $u = u_0 = $ constant. But trigonometry yields

$$\begin{aligned} u &= x \cos \phi_0 + y \sin \phi_0 & x &= u \cos \phi_0 - v \sin \phi_0 \\ v &= -x \sin \phi_0 + y \cos \phi_0 & y &= u \sin \phi_0 + v \cos \phi_0. \end{aligned} \quad (1)$$

So

$$\begin{aligned} R(u_0, \phi_0) &= \int_l f(x, y) \, ds \\ &= \int_{-\infty}^{\infty} f(u_0 \cos \phi_0 - v \sin \phi_0, u_0 \sin \phi_0 + v \cos \phi_0) \, dv. \end{aligned} \quad (2)$$

(Note that if the object has finite extent [so that $f(x, y) = 0$ for (x, y) not in the object], then the above integral actually has finite limits.) The problem is to solve this integral equation for $f(x, y)$

given the ray-sums $R(u, \phi)$. In other words, in a real situation the ray-sums would be known, but the actual mass density would be of interest.

Exercises

1. Derive these equations: Let P be a point with coordinates (x, y) in the x, y coordinate system and (u, v) in the u, v system. Show that (x, y) and (u, v) are related as above.

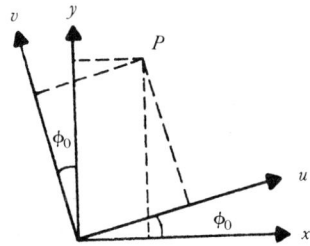

3. Back Projection: An Approximation

Let's first obtain an approximate solution to the problem of reconstructing the unknown mass density $f(x, y)$ from the known ray-sums $R(u, \phi)$. This is the technique of back projection. Suppose in addition to the ray-sums we also know that the object is contained in the disk of radius L about the origin. Now $R(u, \phi)$ is the integral of f along l; and the length of l (or rather the length of the intersection of l with the disk of radius L) is $2\sqrt{L^2 - u^2}$ (see **Figure 5**). Suppose that (x_0, y_0) is a point on l. Under the assumption that

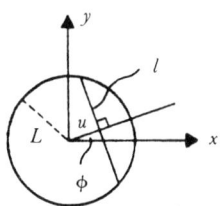

Figure 5. l has length $2\sqrt{L^2 - u^2}$.

all the mass on l is *uniformly* distributed along the line segment l, then the mass density along l would be

$$\delta(x_0, y_0) = \frac{R(u, \phi)}{2\sqrt{L^2 - u^2}}, \qquad (3)$$

where $u = x_0\cos\phi + y_0\sin\phi$ as in **Exercise 2**. Finally we approximate $f(x_0, y_0)$ as the average of all $\delta(x_0, y_0)$ as ϕ varies from 0 to 2π.

Using **Exercise 2** we see that the back projection estimate of f is

$$F(x_0, y_0) = \frac{1}{2\pi} \int_0^{2\pi} \frac{R(x_0\cos\phi + y_0\sin\phi, \phi)}{2\left[L^2 - (x_0\cos\phi + y_0\sin\phi)^2\right]^{\frac{1}{2}}} \, d\phi. \qquad (4)$$

Example: Suppose the object is actually the disk of radius L of uniform density $f(x, y) =$ constant c. Then we would observe the ray-sums (from **(2)**)

$$R(u, \phi) = \begin{cases} 2c\sqrt{L^2 - u^2}, & |u| \le L \\ 0, & |u| > L. \end{cases}$$

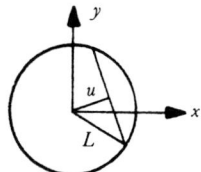

Thus for every u, ϕ

$$\frac{R(u, \phi)}{2\sqrt{L^2 - u^2}} = c \quad \text{for } |u| \le L;$$

and back projection yields approximation

$$F(x_0, y_0) = \frac{1}{2\pi} \int_0^{2\pi} c \, d\phi = c$$

for (x_0, y_0) inside the disk of radius L, which agrees with the exact value of $f(x_0, y_0)$.

In general, the back projection approximation overestimates regions of low mass density and underestimates regions of high mass density. This is due to the fact that projection for each fixed ϕ of

$R(u, \phi)$ *uniformly* back through the object will overestimate regions of low density *and underestimate regions of high density.*

Exercises

2. Show that the equation of line l through (x_0, y_0) at angle ϕ as in **Figure 5** is

$$y = -(\tan \phi)^{-1} x + \left[y_0 + (\tan \phi)^{-1} x_0 \right].$$

Show that the distance from this line to the origin is

$$u = x_0 \cos \phi + y_0 \sin \phi.$$

3. Suppose the object is contained in the disk of radius L about the origin with density $f(x, y) = c(x^2 + y^2)$, c a positive constant. Find the ray-sums $R(u, \phi)$ and the back projection $F(x, y)$. How does F compare with f?

4. Suppose the object is known to be contained in the disk of radius L about the origin, but is actually the disk of radius $L_0 < L$ of constant density c.

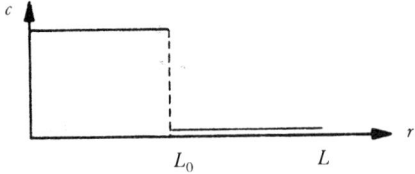

Abscissa represents $r = \sqrt{x^2 + y^2}$ = distance from origin.

What are the ray-sums $R(u, \phi)$ and the back projection approximation $F(x, y)$? (Leave your answer as an integral.) Draw a rough sketch of F as a function of the distance r from the origin.

4. Fourier Transform Notation

Let $f(x)$ be a function of the real variable x, $-\infty < x < \infty$. We use this convention for the *Fourier transform* $\hat{f}$ of f:

$$\hat{f}(\omega) = \int_{-\infty}^{\infty} f(x) e^{-2\pi i \omega x} \, dx.$$

Under suitable conditions the *Fourier inversion theorem* holds

$$f(x) = \int_{-\infty}^{\infty} \hat{f}(\omega) e^{2\pi i \omega x} \, d\omega.$$

(Suitable conditions are, for example, that

$$\int_{-\infty}^{\infty} |f| < \infty, \quad \int_{-\infty}^{\infty} |f|^2 < \infty,$$

f is continuous and is piecewise differentiable.) These results can easily be extended to functions $f(x, y)$ of two variables. Thus we set

$$\hat{f}(\omega, \sigma) = \int_{-\infty}^{\infty} \int_{-\infty}^{\infty} f(x, y) e^{-2\pi i (\omega x + \sigma y)} \, dx \, dy;$$

and under suitable conditions the corresponding Fourier inversion theorem holds:

$$f(x, y) = \int_{-\infty}^{\infty} \int_{-\infty}^{\infty} \hat{f}(\omega, \sigma) e^{2\pi i (\omega x + \sigma y)} \, d\omega \, d\sigma.$$

Exercises

5. Let $f(x, y) = e^{-a(x^2 + y^2)}$, where a is a positive constant. Use contour integration to show that

$$\hat{f}(\omega, \sigma) = \frac{\pi}{a} \exp\left[-\frac{\pi^2}{a}(\omega^2 + \sigma^2)\right].$$

Show that the Fourier inversion theorem holds for this f. (Note: the second part of this problem requires no new integrations.)

5. The Exact Solution

"...the ray-sums... completely determine... the Fourier transform of the mass density."

We now show how the ray-sums $R(u, \phi)$ completely determine $\hat{f}(\omega, \sigma)$, the Fourier transform of the mass density. The trick is to change the variables of integration in the integral defining f above. Switch to the u, v coordinates defined in (1). Orient ω, σ axes—the domain variables for f—so that $\phi = \arctan(\sigma/\omega)$. Then

$$\cos \phi = \sec^{-1}\phi = (1 + \tan^2\phi)^{-1/2}$$
$$= (1 + \sigma^2/\omega^2)^{-1/2} = \frac{\omega}{\sqrt{\omega^2 + \sigma^2}},$$
$$\sin \phi = \frac{\sigma}{\sqrt{\omega^2 + \sigma^2}},$$
$$\omega x + \sigma y = \omega(u \cos \phi - v \sin \phi) + \sigma(u \sin \phi + v \cos \phi)$$
$$= (\sqrt{\omega^2 + \sigma^2}) u.$$

Also $du\,dv = dx\,dy$, since (u, v) and (x, y) are each Cartesian coordinate systems. So

$$\hat{f}(\omega, \sigma) = \int_{-\infty}^{\infty}\int_{-\infty}^{\infty} f(x, y) e^{-2\pi i(\omega x + \sigma y)}\,dx\,dy$$

$$= \int_{-\infty}^{\infty}\int_{-\infty}^{\infty} f(x, y)\,dv \exp\left[-2\pi i\left(\sqrt{\omega^2 + \sigma^2}\right)u\right]\,du, \tag{5}$$

where

$x = u\cos\phi - v\sin\phi$
$y = u\sin\phi + v\cos\phi.$

But the inner integral is $R(u, \phi)$ by (2). Thus

$$\hat{f}(\omega, \sigma) = \int_{-\infty}^{\infty}\int_{-\infty}^{\infty} R(u, \phi)\exp\left[-2\pi i\left(\sqrt{\omega^2 + \sigma^2}\right)u\right]\,du.$$

However, *this* integral is by definition the Fourier transform of $R(u, \phi)$ with respect to the u variable at $\sqrt{\omega^2 + \sigma^2}$ and with ϕ fixed at $\arctan(\sigma/\omega)$. In other words, $f(\omega, \sigma)$ is obtained by fixing ϕ at $\arctan(\sigma/\omega)$, taking the Fourier transform of $R(u, \phi)$ with respect to the single variable u, and evaluating this Fourier transform at $\sqrt{\omega^2 + \sigma^2}$. For a function $g(x, y)$ of two variables, we denote the Fourier transform with respect to the first variable by $\hat{g}_1(\omega, y)$:

$$\hat{g}_1(\omega, y) = \int_{-\infty}^{\infty} g(x, y) e^{-2\pi i\omega x}\,dx.$$

6. The Basic Theorem

Summarizing the above yields the basic theorem:

$$\hat{f}(\omega, \sigma) = \hat{R}_1\left[\sqrt{\omega^2 + \sigma^2}, \arctan\left(\frac{\sigma}{\omega}\right)\right]. \tag{6}$$

The theorem shows us how, in theory, to reconstruct the mass density $f(x, y)$ from the ray-sums $R(u, \phi)$. Namely, from $R(u, \phi)$ calculate the Fourier transform with respect to u, $R_1(\tau, \phi)$. Then $\hat{f}(\omega, \sigma)$ can be calculated by the theorem, from which $f(x, y)$ can be found by Fourier inversion. Of course in practice $R(u, \phi)$ is much too complicated to find its Fourier transform without a computer approximation of the integral defining $\hat{R}_1(\tau, \sigma)$, and similarly for the step from $\hat{f}$ to f. Still, the theorem indicates the general technique.

Example: Suppose the ray-sums are found to be

$$R(u, \phi) = e^{-au^2}$$

for positive constant a. (Noting that $R(u, \phi)$ is independent of ϕ, we expect $f(x, y)$ to be a function of the distance $x^2 + y^2$ from the origin only.) Now

$$\hat{R}_1(\tau, \phi) = \int_{-\infty}^{\infty} e^{-au^2} e^{-2\pi i \tau u} \, du$$

$$= \sqrt{\frac{\pi}{a}} \exp\left[-\frac{\pi^2}{a}\tau^2\right]$$

either using tables or by a contour integration as in **Exercise 5**. So

$$\hat{f}(\omega, \sigma) = \hat{R}_1(\sqrt{\omega^2 + \sigma^2}, \arctan(\sigma/\omega))$$

$$= \sqrt{\frac{\pi}{a}} \exp\left[-\frac{\pi^2}{a}(\omega^2 + \sigma^2)\right].$$

Thus

$$f(x, y) = \int_{-\infty}^{\infty}\int_{-\infty}^{\infty} \hat{f}(\omega, \sigma) e^{2\pi i(\omega x + \sigma y)} \, d\omega \, d\sigma$$

$$= \sqrt{\frac{\pi}{a}} \int_{-\infty}^{\infty} \exp\left[-\frac{\pi^2}{a}\omega^2 + 2\pi i \omega x\right] d\omega$$

$$\cdot \int_{-\infty}^{\infty} \exp\left[-\frac{\pi^2}{a}\sigma^2 + 2\pi i \sigma y\right] d\sigma$$

$$= \sqrt{\frac{\pi}{a}} \cdot \sqrt{\frac{a}{\pi}} e^{-ax^2} \cdot \sqrt{\frac{a}{\pi}} e^{-ay^2}$$

$$= \sqrt{\frac{a}{\pi}} \exp[-a(x^2 + y^2)]$$

where the second to the last equality follows by a contour integration as in **Exercise 5**.

Exercises
6. Suppose $R(u, \phi)$ is found to be $e^{-|u|}$ independently of ϕ. Find $\hat{R}_1(\tau, \sigma)$, $\hat{f}(\omega, \sigma)$, set up the integral for $f(x, y)$.

7. Circular Symmetry

Many corollaries follow from the basic theorem of (6). We indicate one that generalizes **Exercise 6** and the example before it.

Suppose $R(u, \phi)$ exhibits some sort of periodic behavior in ϕ. That is, suppose the x-ray photographs begin to repeat themselves as we move the angle of the incident x-ray beam. (Of course such a repetition always occurs by moving the beam by 2π. We have smaller periods of repetition in mind here.) Then it is natural to use polar rather than Cartesian coordinates. Thus let

$$x = r \cos \theta$$
$$y = r \sin \theta,$$

as always for polar coordinates. In the (ω, σ) plane let

$$\omega = t \cos \beta$$
$$\sigma = t \sin \beta$$

denote the polar representation. That is,

$$t = \sqrt{\omega^2 + \sigma^2}, \quad \beta = \arctan(\sigma/\omega).$$

Now

$$f(x, y) = \int_{-\infty}^{\infty}\int_{-\infty}^{\infty} \hat{f}(\omega, \sigma) e^{2\pi i(\omega x + \sigma y)} d\omega \, d\sigma$$

$$= \int_{-\infty}^{\infty}\int_{-\infty}^{\infty} \hat{R}_1\left[\sqrt{\omega^2 + \sigma^2}, \arctan(\sigma/\omega)\right] e^{2\pi i(\omega x + \sigma y)} d\omega \, d\sigma$$

$$= \int_{0}^{\infty}\int_{0}^{2\pi} \hat{R}_1(t, \beta) e^{2\pi i rt \cos(\beta - \theta)} t d\beta \, dt, \qquad (7)$$

where the second equality follows from the basic theorem, and the last by using polar coordinates specifically, the fact that

$$\omega x + \sigma y = rt \cos \beta \cos \theta + rt \sin \beta \sin \theta$$
$$= rt \cos(\beta - \theta).$$

Note that $td\beta \, dt$ is the area element in polar coordinates. By **Exercise 7** below we conclude that

$$f(r, \theta) = \int_{0}^{\infty}\int_{0}^{2\pi} \hat{R}_1(t, \beta + \theta) e^{2\pi i rt \cos \beta} t d\beta \, dt. \qquad (8)$$

So far we have been completely general. To understand the above equation more, let's take the case $R(u, \phi) = R(u)$ independently of ϕ. Then $\hat{R}_1(\tau, \phi) = \hat{R}_1(\tau)$ is independent of ϕ (why?). So

$$f(r, \theta) = \int_0^\infty \int_0^{2\pi} \hat{R}_1(t) e^{2\pi i r t \cos \beta} t \, d\beta \, dt. \tag{9}$$

Noting that the integral is independent of θ, we have just proved the intuitive feature: If the ray-sums $R(u, \phi)$ are independent of the angle of the incident x-rays, then f must be circularly symmetric, i.e., independent of θ.

Using **Exercise 8** and its solution, we see that if $R(u, \phi)$ is independent of ϕ, then (from **(9)**)

$$f(r, \theta) = f(r)$$

$$= \int_0^\infty \hat{R}_1(t) \left[\int_0^{2\pi} e^{2\pi i r t \cos \beta} \, d\beta \right] t \, dt$$

$$= \int_0^\infty \hat{R}_1(t) \cdot 2\pi J_0(2\pi r t) t \, dt, \tag{10}$$

where $J_0(\lambda)$ is the Bessel function of order zero

$$J_0(\lambda) = \sum_{j=0}^\infty \frac{(-1)^j}{(j!)^2} \left(\frac{\lambda}{2} \right)^{2j}.$$

(See the answer to **Exercise 8**.) This formula exhibits $f = f(r)$ as what is called the *Hankel transform* of order zero of $\hat{R}_1(t)$. Hankel transforms occur quite naturally in two-dimensional Fourier analysis problems.

Although this formula exhibiting $f(r)$ as the Hankel transform of $\hat{R}_1(t)$ solves the case when $R(u, \phi) = R(u)$ is independent of ϕ, the actual integration would normally require a computer approximation. However, the Hankel transforms of some common functions are listed in mathematical tables.

Exercises

7. Let g be a real-valued function of a real variable which is periodic of period 2:

$$g(\beta + 2\pi) = g(\beta)$$

for all β. Let c be a fixed number. By considering the graph of g, show that

$$\int_c^{2\pi + c} g(\beta) \, d\beta = \int_0^{2\pi} g(\beta) \, d\beta.$$

Noting that both $\cos\beta$ and $\hat{R}_1(t, \beta + \theta)$ are periodic of period 2π (so that their product is as well), derive (8) from the last expression in (7).

8. Let λ be a real constant. The value of

$$\int_0^{2\pi} e^{i\lambda \cos s}\, ds$$

is denoted $2\pi J_0(\lambda)$. (J_0 is called the *Bessel function of order zero*.) Evaluate this integral as an infinite sum using contour integration. (Hint: Convert to a contour integral over the counterclockwise unit circle centered at the origin using $z = e^{is}$. Show that the integral is

$$\int_C \exp\left[i\frac{\lambda}{2}\left(z + \frac{1}{z}\right)\right]\frac{dz}{iz}$$

which is $(2\pi i)$ Residue of $\dfrac{1}{iz}\exp[i(\lambda/2)(z + (1/z))]$ at $z = 0$. Now evaluate this residue as an infinite series.)

9. Suppose the ray-sums are found to be independent of ϕ with

$$R(u, \phi) = R(u) = \frac{2}{1 + (2\pi u)^2}.$$

By a contour integration show that $\hat{R}_1(t) = e^{-|t|}$. Show that

$$\int_0^\infty e^{-t} \cdot t^{2j+1}\, dt = (2j + 1)!$$

and that $f(r, \theta) = 2\pi(1 + (\pi r)^2)^{-3/2}$. (Note: You will need the fact that

$$\binom{2j}{j} = \binom{-1/2}{j}(-1)^j$$

and the binomial theorem in the special case

$$(1 + x)^{-1/2} = \sum_{j=0}^\infty \binom{-1/2}{j} x^j,$$

where by definition

$$\binom{-1/2}{j} = \frac{(-1/2)(-1/2 - 1)\cdots(-1/2 - j + 1)}{j!}.)$$

8. The Circularly Asymmetric Case

A similar analysis can be carried out when $R(u,\phi)$ does depend on ϕ. The solution exhibits $f(r,\theta)$ as combinations of higher-order Hankel transforms. The solution is elegant, but we relegate it to **Exercise 10**.

Exercises

10. Now $\hat{R}_1(t,\phi)$—the Fourier transform of $R(u,\phi)$ with respect to the u variable—is for each fixed t a periodic function of ϕ of period 2π. This is so since the ray-sums $R(u,\phi)$ repeat as ϕ changes by 2π. Thus $\hat{R}_1(t,\phi)$ can be expanded as a Fourier series for each fixed t:

$$\hat{R}_1(t,\phi) = \sum_{n=-\infty}^{\infty} \hat{c}_n(t) e^{in\phi},$$

where by the Fourier series formula

$$\hat{c}_n(t) = \frac{1}{2\pi} \int_0^{2\pi} \hat{R}_1(t,\phi) e^{-in\phi}\, d\phi.$$

a. Show that $\hat{c}_n(t)$ is the Fourier transform of the nth Fourier coefficient of $R(u,\phi)$

$$c_n(u) = \frac{1}{2\pi} \int_0^{2\pi} R(u,\phi) e^{-in\phi}\, d\phi.$$

b. Plus the new expression for $\hat{R}_1(t,\phi)$ into (8) to show that

$$f(r,\theta) = \sum_{n=-\infty}^{\infty} e^{in\theta} \cdot \int_0^{\infty} \hat{c}_n(t) d_n(rt) t\, dt,$$

where

$$d_n(rt) = \int_0^{2\pi} \exp[ins + 2\pi irt \cdot \cos s]\, ds.$$

By definition $J_n(\lambda) = (2\pi i^n)^{-1} d_n(\lambda/2\pi)$ is called the *Bessel function of order n*.

c. Now fix r and consider $f(r,\theta)$ as a function of θ. Conclude from **Exercise 10a** and **10b** above, that the nth Fourier coefficient of $f(r, \theta)$ when expanded in a Fourier series as a function of θ is

$$2\pi i^n \int_0^\infty \hat{c}_n(t) J_n(2\pi rt) t \, dt,$$

which is i^n times the nth Hankel transform of $\hat{c}_n$.

9. Historical and Bibliographical Note

Analytic techniques in tomography were first used in radio-astronomy. Antennae were unable to focus on the individual points of the solar surface but were able to focus on thin striplike segments crossing the sun. Thus, in making a map of the microwave radiation density of the sun, astronomers had to reconstruct it from "strip sums." In medical radiology, x-ray CAT (computer-assisted tomography) scanners were introduced in 1972. These machines take x-ray pictures at about one degree intervals between $0°$ and $180°$. While CAT scanners are expensive (on the order of half a million dollars per machine), they afford a revolutionary improvement in medical diagnosis.

For an interesting non-technical source of information see "Computerized tomography" by W. Swindell and H. Barrett in *Physics Today* (December 1977). For a deeper mathematical treatment, see "Principles of computer assisted tomography" by R. Brooks and G. DiChiro in *Physics in Medicine and Biology* 21 (1976): 689–732. For an applied, very readable account of Fourier series, Fourier transforms and Hankel transforms, see *The Fourier Transform and its Applications* by R. M. Bracewell (McGraw-Hill, 1965). An informative, nonmathematical article on a similar method of image reconstruction is "Ultrasound in medical diagnosis" by G. Devey and P. Wells in *Scientific American* (May 1978): waves 98–112. In this technique, sound rather than x-rays are used, in order to avoid organ damage.

10. Answers and Suggestions to Exercises

3. $R(u, \phi) = \dfrac{2c}{3}\sqrt{L^2 - u^2}\,(L^2 + 2u^2)$ from **(2)**.

$F(x_0, y_0) = \dfrac{c}{3}(L^2 + x_0^2 + y_0^2)$ from **(4)**.

As functions of $r = \sqrt{x^2 + y^2}$, the graphs of f and F are

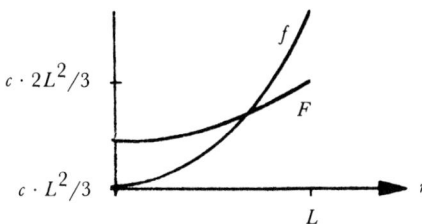

4. $R(u, \phi) = \begin{cases} 2c\sqrt{L_0^2 - u^2}, & |u| \leq L_0 \\ 0, & L_0 < |u| \leq L. \end{cases}$

So by **(2)**

$$F(x_0, y_0) = \frac{1}{2\pi} \int c \sqrt{\frac{L_0^2 - u^2}{L^2 - u^2}} \, d\phi,$$

where $u = x_0 \cos \phi + y_0 \sin \phi$ and the range of integration is over all $0 \leq \phi \leq 2\pi$ so that $|u| \leq L_0$. Note that

$$F(0, 0) = \frac{cL_0}{L}$$

$$F(x_0, y_0) = 0$$

for $\sqrt{x_0^2 + y_0^2} = L$. A rough sketch of F as a function of the distance from the origin is

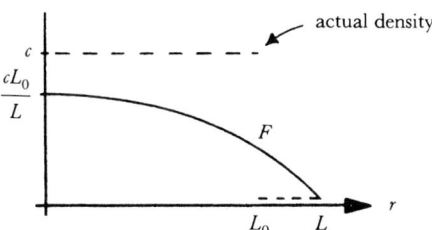

5. $\hat{f}(\omega, \sigma) = \iint e^{-2\pi i(\omega x + \sigma y)} \cdot e^{-a(x^2 + y^2)} \, dx \, dy$

$= \int e^{-2\pi i \omega x} e^{-ax^2} \, dx \cdot \int e^{-2\pi i \sigma y} e^{-ay^2} \, dy.$

But the one-dimensional Fourier Transform of e^{-ax^2} is $\sqrt{\pi/a} \exp[-(\pi^2/a)\omega^2]$, from tables or by contour integration. Now with

$$\hat{f}(\omega, \sigma) = (\pi/a)\exp[-(\pi^2/a)(\omega^2 + \sigma^2)],$$

$$\hat{\hat{f}}(x, y) = \frac{\pi}{a} \cdot \frac{\pi}{\pi^2/a} \exp\left[-\frac{\pi^2}{\pi^2/a}(x^2 + y^2)\right]$$

$$= e^{-a(x^2+y^2)}$$

by the above integration with π^2/a replacing a. So

$$\hat{\hat{f}}(-x, -y) = \hat{\hat{f}}(x, y) = f(x, y)$$

which is the content of the Fourier Inversion Theorem.

6. $\hat{R}_1(\tau, \phi) = 2(1 + 4\pi^2\tau^2)^{-1}$. Thus

$$\hat{f}(\omega, \sigma) = 2[1 + 4\pi^2(\omega^2 + \sigma^2)]^{-1}.$$

8. On the unit circle

$$z = e^{is}, \quad \cos s = \frac{1}{2}\left(z + \frac{1}{z}\right), \quad \frac{dz}{iz} = ds.$$

Now

$$e^{a(z+1/z)} = \sum_{j=0}^{\infty} \frac{(az)^j}{j!} \cdot \sum_{k=0}^{\infty} \frac{(az^{-1})^k}{k!}.$$

The coefficient of z^0 in the product series is

$$\sum_{j=0}^{\infty} \frac{a^j \, a^j}{j! \, j!} = \sum_{j=0}^{\infty} \frac{a^{2j}}{(j!)^2}.$$

Thus the coefficient of z^{-1} in $(1/iz)\exp[i(\lambda/2)(z + (1/z))]$ is

$$\frac{1}{i}\sum_{j=0}^{\infty}\left(\frac{i\lambda}{2}\right)^{2j} \cdot \frac{1}{(j!)^2} = \frac{1}{i}\sum_{j=0}^{\infty}\frac{(-1)^j}{(j!)^2}\left(\frac{\lambda}{2}\right)^{2j}.$$

$2\pi i$ times this is the answer by the residue theorem.

9. With $\hat{R}(t) = e^{-|t|}$, the formulas above the statement of this problem imply

$$f(r, \theta) = 2\pi\int_0^\infty e^{-t} \cdot \sum_{j=0}^\infty \frac{(-1)^j}{(j!)^2}(\pi r)^{2j} t^{2j+1} \, dt$$

$$= 2\pi \sum_{j=0}^\infty \frac{(-1)^j}{(j!)^2}(\pi r)^{2j} \cdot (2j+1)!$$

$$= 2\pi \sum \binom{2j}{j}(-1)^j(2j+1)(\pi r)^{2j}$$

$$= 2\pi\frac{d}{d(\pi r)}\sum\binom{2j}{j}(-1)^j(\pi r)^{2j+1}$$

$$= 2\pi\frac{d}{d(\pi r)}\pi r\sum\binom{-1/2}{j}((\pi r)^2)^j$$

$$= 2\pi\frac{d}{d(\pi r)}\frac{\pi r}{\sqrt{1 + (\pi r)^2}}.$$

UMAP

Modules in Undergraduate Mathematics and its Applications

Module 474

Nominal vs. Effective Rates of Interest

Peter A. Lindstrom

Published in cooperation with the Society for Industrial and Applied Mathematics, the Mathematical Association of America, the National Council of Teachers of Mathematics, the American Mathematical Association of Two-Year Colleges, The Institute of Management Sciences, and the American Statistical Association.

A word of warning about Money Market Accounts: Carefully consider a premature withdrawl of your investment.

By permission of Johnny Hart and Field Enterprizes, Inc.

INTERMODULAR DESCRIPTION SHEET:	UMAP Unit 474
TITLE:	Nominal vs. Effective Rates of Interest
AUTHOR:	Peter A. Lindstrom Math/Tech Division North Lake College Irving, TX 75038
MATHEMATICS FIELD:	Mathematics of Finance
APPLICATIONS FIELD:	Banking, business, and economics
ABSTRACT:	In reading newspapers and magazines, in listening to the radio, and in watching television, one is often confronted with the terms "compounded daily," "the effective annual yield," "the rate of inflation for last month was...," etc. Such items are applications of two forms of interest, the *nominal rate* and the *effective rate*. The purpose of this module is to 1) introduce these terms, 2) show how they are used in banking practices, in discussing the rate of inflation, and in other aspects of everyone's life, and 3) show how these terms are often misused in advertising and by the media. Students will develop an understanding of the terms "nominal rate" and "effective rate" and learn how these terms are used in the everyday life of a consumer.
PREREQUISITES:	Percentages, exponents, solving equations, and how to use a hand calculator.

© Copyright 1983, 1988 COMAP, Inc. All rights reserved.

COMAP, 60 Lowell Street, Arlington, MA 02174 (617) 641-2600

Nominal vs. Effective Rates of Interest

Peter A. Lindstrom
Math/Tech Division
North Lake College
Irving, TX 75038

Table of Contents

1. INTRODUCTION 1
2. NOMINAL RATE AND EFFECTIVE RATE 1
   2.1 Definitions, Formulas, and Review 1
   2.2 Effective Rates for Comparison 5
   2.3 Compounding Daily and Effective Rate 7
   2.4 Compounding Continuously and Effective Rate 13
   2.5 Nominal and Effective Rates of Inflation 18
   2.6 Nominal and Effective Rates in Other Areas 20
3. CONCLUSION 23
4. MODEL EXAM 23
5. ANSWERS TO EXERCISES 25
6. ANSWERS TO MODEL EXAM 28

MODULES AND MONOGRAPHS IN UNDERGRADUATE
MATHEMATICS AND ITS APPLICATIONS (UMAP) PROJECT

The goal of UMAP was to develop, through a community of users
developers, a system of instructional modules in undergraduate mathem
and its applications to be used to supplement existing courses and f
which complete courses may eventually be built.

The Project was guided by a National Advisory Board of mathen
cians, scientists, and educators. UMAP was funded by a grant from
National Science Foundation and is now supported by the Consortium
Mathematics and Its Applications (COMAP), Inc., a non-profit corpora
engaged in research and development in mathematics education.

COMAP STAFF

Paul J. Campbell	Editor
Solomon A. Garfunkel	Executive Director, COMAP
Laurie W. Aragon	Business Development Manager
Philip A. McGaw	Production Manager
Theresa Cronin	Copy Editor
Annemarie S. Morgan	Administrative Assistant
John Gately	Distribution

This unit was developed under the auspices of the UMAP Precalculus Panel whose members were: Peter A. Lindstrom (Chair), North Lake College; Charles Biles, Humboldt State University; Marilyn Mays, North Lake College; Roland Lamberson, Humboldt State University; and Richard G. Montgomery, Southern Oregon State College. The Project would like to thank Dennis R. Adams of the Tennessee State Technical Institute, Martha Kasting of the University of Louisville, Joel Greenstein of New York City Technical College, Harold Baker of Litchfield High School, and Gerald Egerer of Sonoma State University for their reviews, and all others who assisted in the production of this unit.

This material was prepared with the partial support of National Science Foundation Grants No. SED80-07731 and No. SPE-8304192. Recommendations expressed are those of the author and do not necessarily reflect the views of the NSF or the copyright holder.

1. Introduction

In reading newspapers and magazines, in listening to the radio, and in watching television, one is often confronted with the terms "compounded daily," "the effective annual yield," "the rate of inflation for last month was...," etc. Such items are applications of two forms of interest, the *nominal rate* and the *effective rate*. The purpose of this module is to (1) introduce these terms; (2) show how they are used in banking practices, in discussing the rate of inflation, and in other aspects of everyone's life; and (3) show how these terms are often misused in advertising and by the media.

"...a calculator with an 'exponential key' is essential."

As you work through this module, a hand-held calculator will be necessary. Since exponents are involved in many of the calculations, a calculator with an "exponential key" is essential. A Texas Instrument SR-50A was used in developing this module; should you use a different calculator, some of your results might differ slightly from what you find here.

2. Nominal Rate and Effective Rate

2.1. Definitions, Formulas, and Review

A person who invests (borrows) money, receives (pays) *interest* on the investment (loan). The money invested (borrowed) is called the *principal*. The sum of the principal and the interest is called the *amount*. The *rate of interest* is expressed as a percent of the principal for a specified period of time, where time is measured in years. The rate of interest is expressed as a yearly rate. Throughout this module, two different types of yearly interest will be discussed.

  a. **Simple Interest**: *Simple interest* is interest paid only on the principal invested (borrowed). For simple interest, the interest period is the length of time for the investment (loan).

  b. **Compound Interest**: When the length of time for an investment (loan) is divided into shorter time periods of equal length and when the interest for each time period is added to the principal to obtain a new principal for computing the interest for the next period, the interest is called *compound interest*.

The basic formula for computing simple interest is:

Interest = (Principal) × (Simple interest rate as a decimal)
× (time expressed in years)

or

$$I = Prt. \tag{A}$$

As you will soon see, **(A)** can be used for developing other formulas for both simple interest problems and compound interest problems.

Example 1.
Suppose that you invest $1,000 today for two years. Determine the interest and the amount of your investment if (a) the rate of interest is 8% simple interest and (b) the rate of interest is 8% compounded semiannually.

Solution:
a. For 8% simple interest, **(A)** becomes

$$I = (\$1{,}000)(.08)(2),$$
$$= \$160.$$

Since the amount = principal + interest, then

$$\text{amount} = \$1{,}000 + 160,$$
$$= \$1{,}160.$$

Before solving (b), let's develop another formula for the amount S:

$$S = P + I$$
$$= P + Prt$$
$$= P(1 + rt). \tag{B}$$

Notice that **(B)** yields the amount S, if you know the principal P, the simple interest rate r, and the length of time t expressed in years. Using **(B)** to solve part (a) again, you have

$$S = \$1{,}000(1 + (.08)(2)) = \$1{,}160.$$

b. To solve this problem, recall that the 8% (compounded semiannually) is a yearly rate and the total length of the investment is 2 years or 4 semiannual periods. In **(B)**, the expression $rt = (.08)(1/2) = .04$ for each semiannual interest period. Considering each half-year interest period, we then have the following amounts:

$$\text{Amount for first half-year} = S_1 = \$1{,}000(1 + .04)$$
$$= \$1{,}000(1.04)$$
$$= \$1040.$$
$$\text{Amount for second half-year} = S_2 = S_1(1 + .04)$$
$$= \$1040(1.04)$$
$$= \$1081.60.$$
$$\text{Amount for third half-year} = S_3 = S_2(1 + .04)$$
$$= \$1081.60(1.04)$$
$$= \$1124.86.$$
$$\text{Amount for fourth half-year} = S_4 = S_3(1 + .04)$$
$$= \$1124.86(1.04)$$
$$= \$1169.85.$$

Hence, you will have \$1169.85 for the amount of your investment after two years, and the interest will be $\$1169.85 - \$1{,}000 = \$169.85$.

The interest and the amount in **(b)** are both larger than the corresponding items in **(a)**. This should not be surprising, for in **(b)** both the principal and the interest are earning interest, whereas in **(a)** only the principal is earning the interest.

Tracing backward through the computations that produced S_4, notice that:

$$S_4 = S_3(1 + .04),$$
$$= S_2(1 + .04)(1 + .04),$$
$$\text{since } S_3 = S_2(1 + .04),$$
$$= S_1(1 + .04)(1 + .04)(1 + .04),$$
$$\text{since } S_2 = S_1(1 + .04),$$
$$= \$1{,}000(1 + .04)(1 + .04)(1 + .04)(1 + .04),$$
$$\text{since } S_1 = \$1{,}000(1 + .04),$$
$$= \$1{,}000(1 + .04)^4,$$

or, $S_4 = \underbrace{\$1{,}000}_{\text{principal}}\underbrace{(1 + .08/2)}_{\substack{\text{rate per}\\\text{half-year}\\\text{interest}\\\text{period}}}{}^{\overset{\displaystyle 4}{\nwarrow\text{the number of half-year}\atop\text{interest periods}}}$

This expression can be generalized to the following:

$$\underset{\text{amount}}{S} = \underset{\text{principal}}{P}(1 + \underset{\substack{\text{rate per}\\\text{interest}\\\text{period}}}{i})^{\overset{\displaystyle n}{\searrow\text{number of}\atop\text{interest periods}}} \qquad (\text{C})$$

Formula **(C)** will be used throughout this module.

Exercise
1. **a.** Invest $750 today for 18 months at 7.5% simple interest rate. Determine the interest and the amount of this transaction.
   b. Redo **Exercise 1(a)** using 7.5% compounded quarterly.

We now come to the two terms that serve as the basis of this module:

Nominal rate: The *nominal rate*, or the *nominal annual rate*, is that *yearly rate* of interest that is compounded *more than once per year*. This rate is also called the *annual interest rate*.

Effective rate: The *effective rate* or the *effective annual rate* is that *yearly rate* of interest that is compounded *only once per year*. This rate is also called the *annual equivalent yield*.

In **Example 1(b)**, the 8% compounded semiannually is a *nominal rate*, as is the 7.5% compounded quarterly found in **Exercise 1(b)**. The rates in **Example 1(a)** and **Exercise 1(a)** are both simple interest rates. (Why is neither one of these an effective rate?) If you invest your money at 10.5% compounded annually, then the 10.5% is an effective rate.

One last item to consider is the following:

Equivalent rates: Two yearly rates of interest are said to be *equivalent* if both rates produce the same amount *in one year*.

Exercise

2. a. Determine the amount of each of the following investments:
   (i) invest $500 for one year at 10.5% effective rate and
   (ii) invest $500 at a 10.5% simple interest rate for one year.

   b. Show that the *effective rate* is *equivalent* to the *simple interest rate for one year*.

2.2 Effective Rates for Comparison

In 1969 the U.S. Federal Government passed the Truth-in-Lending Law. One of the purposes of this law is to let you, the consumer, know exactly what the interest rate is for purposes of making comparisons. By having comparative rates, consumers are able to decide which rate is best for them.

Example 2.
If you were to invest $1,000 today for three years, which of the following *nominal rates* would give you the most money for your investment: 7.85% compounded monthly or 7.9% compounded semiannually?

Solution:
At the 7.85% compounded monthly rate, after three years you will have

$$S = P(1 + i)^n = \$1{,}000(1 + .0785/12)^{36} = \$1264.57.$$

At the 7.9% compounded semiannually rate, after three years you will have

$$S = P(1 + i)^n = \$1{,}000(1 + .079/2)^6 = \$1261.67.$$

Hence, the above calculations indicate that the 7.85% compounded monthly rate yields more money than does the 7.9% compounded semiannually rate.

Example 2 shows that the *smaller* numerical rate produces more money than the *larger* numerical rate. To some consumers this might be surprising as they often feel that the larger the rate, the more money they will have. **Example 2** shows you the following: Do *not* compare the given numerical values of the rates; compare the amounts of money that you would have at the end of the investment period.

Another method of comparing rates is to compare the *effective rates* for each of these nominal rates; i.e., find the effective rate that is *equivalent* to each of the given nominal rates. Returning to **Example 2**, let r_1 be the effective rate that is equivalent to the nominal rate of 7.85% compounded monthly. Since both rates will produce the same amount of money in one year, we have the following equation to solve for r_1:

$$\$1{,}000(1 + r_1/1)^1 = \$1{,}000(1 + .0785/12)^{12}$$
$$(1 + r_1) = (1 + .0785/12)^{12}$$
$$r_1 = (1 + .0785/12)^{12} - 1$$
$$= .0814 \quad \text{or} \quad 8.14\%. \tag{D}$$

Hence, $r_1 = 8.14\%$ is the effective rate that is equivalent to 7.85% compounded monthly.

Let r_2 be the effective rate that is equivalent to the nominal rate of 7.9% compounded semiannually. Again, since both rates will produce the same amount of money in one year, we have the following equation to solve for r_2:

$$\$1{,}000(1 + r_2/1)^1 = \$1{,}000(1 + .079/2)^2$$
$$(1 + r_2) = (1 + .079/2)^2$$
$$r_2 = (1 + .079/2)^2 - 1$$
$$= .0806 \quad \text{or} \quad 8.06\%. \tag{E}$$

Hence, $r_2 = 8.06\%$ is the effective rate that is equivalent to 7.9% compounded semiannually.

You now have two rates, r_1 and r_2, to compare that have the same conversion period; both are compounded only once a year. Since $r_1 > r_2$, after one year you will have more money with the 8.14% effective rate than with the 8.06% effective rate. In terms of the equivalent nominal rates, the 7.85% compounded monthly will produce more money than the 7.9% compounded semiannually.

Items **(D)** and **(E)** can be generalized to determine the effective rate $r\%$ that is equivalent to a nominal rate of $j\%$ compounded k times per year, where

$$r = (1 + j/k)^k - 1.$$

Exercises

3. By comparing *effective rates*, which of the following *nominal rates* is the best for you when you invest your money?
   a. 9.51% compounded weekly.
   b. 9.52% compounded monthly.
   c. 9.53% compounded quarterly.

4. a. In this section, you saw that 7.85% compounded monthly is equivalent to 8.14% compounded yearly. Using the 8.14% effective rate, find the amount you have after 3 years if $1,000 is invested today. Compare the result with that obtained in **Example 2**.
  b. Redo (a) using the 8.06% effective rate that is equivalent to 7.9% compounded semiannually. (This fact was also established in this section.)

2.3 Compounding Daily and Effective Rate

We have just seen that you should compare effective rates and *not* nominal rates in order to determine the best rate for investing your money. By the Truth-in-Lending Law, banks are required to post both the *nominal rate* and the *effective rate*. These rates are often called the *annual rate* and the *annual yield* respectively. **Figure 1** shows two bank advertisements that use these terms.

Notice that in comparing the two advertisements in **Figure 1**, the same nominal rates (annual rates) do *not* produce the same effective rates (annual yields). For example, in one ad, the nominal

PAL SAVINGS & LOAN

ANNUAL RATE		ANNUAL YIELD
8.00%	8-Year $100 minimum	8.33%
7.75%	6-Year $100 minimum	8.06%
7.50%	4-Year $100 minimum	7.79%
*2½-Year Treasury Rate Certificate		
10.40%	$500 minimum	10.96%
*2½-Year Treasury Rate Certificate. Effective Jan. 1 thru Jan 31. The offered rate changes monthly and is determined by the U.S. Treasury. Once this type certificate is issued the rate does not change over the term of the certificate.		
6.50%	1-Year $100 minimum	6.72%
6.00%	3-Month $100 minimum	6.18%
Penalty for early withdrawal of certificates.		
5.50%	Passbook Savings $25 Minimum	5.65%
Paid from date of deposit to date of withdrawal - no penalty.		

Figure 1a.

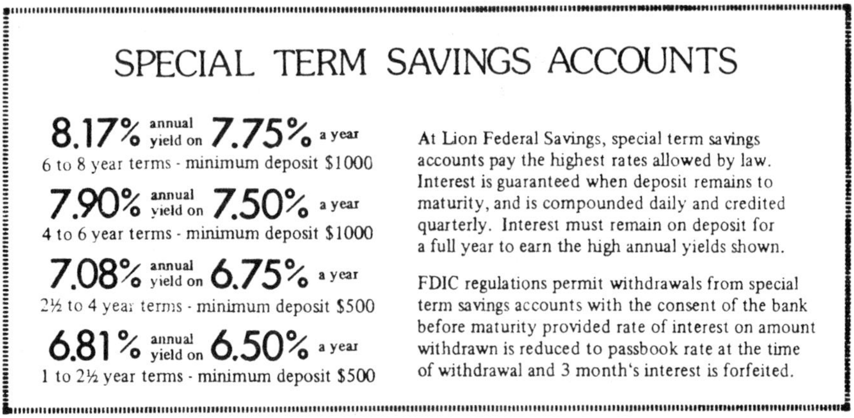

SPECIAL TERM SAVINGS ACCOUNTS

8.17% annual yield on **7.75%** a year
6 to 8 year terms · minimum deposit $1000

7.90% annual yield on **7.50%** a year
4 to 6 year terms · minimum deposit $1000

7.08% annual yield on **6.75%** a year
2½ to 4 year terms · minimum deposit $500

6.81% annual yield on **6.50%** a year
1 to 2½ year terms · minimum deposit $500

At Lion Federal Savings, special term savings accounts pay the highest rates allowed by law. Interest is guaranteed when deposit remains to maturity, and is compounded daily and credited quarterly. Interest must remain on deposit for a full year to earn the high annual yields shown.

FDIC regulations permit withdrawals from special term savings accounts with the consent of the bank before maturity provided rate of interest on amount withdrawn is reduced to passbook rate at the time of withdrawal and 3 month's interest is forfeited.

Figure 1b.

"...some banks use a year with 360 days, while others use 365 days."

rate is 7.75% and its equivalent effective rate is 8.17%, while in the other, the same nominal rate of 7.75% has an effective rate of 8.06%. The reason for this is the manner in which interest is *compounded daily*. When the nominal rate is compounded daily, some banks use a year with 360 days, while others use 365 days.

Example 3.
Let's now consider four possible ways of determining an effective rate r for a nominal rate of 7.75% compounded daily.

Solution:
Case 1: A year with 360 interest periods and a daily rate of 7.75%/360.

$$P(1 + r/1)^1 = P(1 + .0775/360)^{360}$$

$$1 + r = (1 + .0775/360)^{360}$$

$$r = (1 + .0775/360)^{360} - 1$$

$$r = .0806 \quad \text{or} \quad r = 8.06\%.$$

This is called the "360 over 360" method.

8

Case 2: A year with 365 interest periods and a daily rate of 7.75%/365.

$$P(1 + r/1)^1 = P(1 + .0775/365)^{365}$$
$$1 + r = (1 + .0775/365)^{365}$$
$$r = (1 + .0775/365)^{365} - 1$$
$$r = .0806 \quad \text{or} \quad r = 8.06\%.$$

This is called the "365 over 365" method.

Case 3: A year with 365 interest periods and a daily rate of 7.75%/360.

$$P(1 + r/1)^1 = P(1 + .0775/360)^{365}$$
$$1 + r = (1 + .0775/360)^{365}$$
$$r = (1 + .0775/360)^{365} - 1$$
$$r = .0817 \quad \text{or} \quad r = 8.17\%.$$

This is called the "365 over 360" method.

Case 4: A year with 360 interest periods and a daily rate of 7.75%/365.

$$P(1 + r/1)^1 = P(1 + .0775/365)^{360}$$
$$1 + r = (1 + .0775/365)^{360}$$
$$r = (1 + .0775/365)^{360} - 1$$
$$r = .0794 \quad \text{or} \quad r = 7.94\%.$$

This is called the "360 over 365" method.

Case 3, the "365 over 360" method, shows that the "best" effective rate for the investor is obtained when the daily rate is $j\%/360$ and the interest is compounded 365 times per year. All nominal rates *compounded daily* in this module are done by the "365 over 360" method, or "366 over 360" in leap years, unless stated otherwise.

Exercises

5. Verify the other effective rates in **Figure 1**.

6. In **Figure 2**, the 12% nominal rate compounded daily produces effective annual yields of 12.94% and 12.747%. Determine which of the four cases above produce these different effective rates.

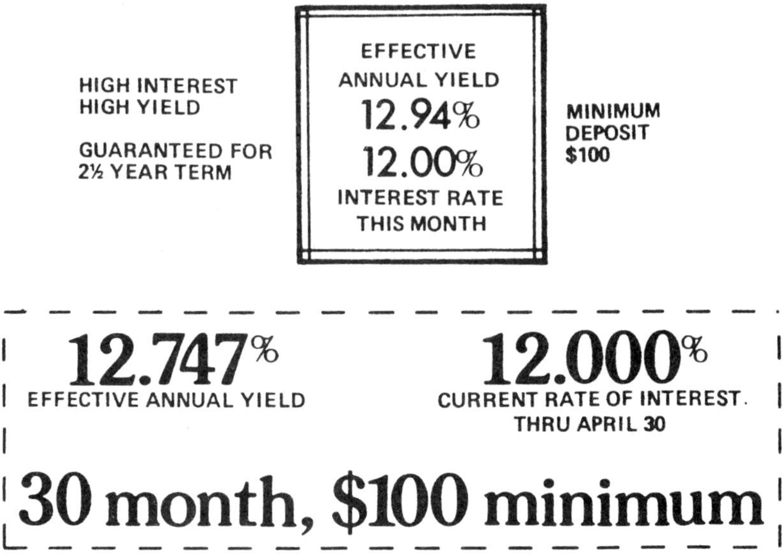

```
                    ┌──────────────────┐
                    │    EFFECTIVE     │
    HIGH INTEREST   │  ANNUAL YIELD    │
    HIGH YIELD      │     12.94%       │   MINIMUM
                    │                  │   DEPOSIT
    GUARANTEED FOR  │     12.00%       │   $100
    2½ YEAR TERM    │   INTEREST RATE  │
                    │    THIS MONTH    │
                    └──────────────────┘
```

12.747%   **12.000%**
EFFECTIVE ANNUAL YIELD   CURRENT RATE OF INTEREST.
                          THRU APRIL 30

30 month, $100 minimum

Figure 2.

7. **a.** Verify that 11.15% is the effective rate that corresponds to a 10.40% nominal rate compounded daily as shown in **Figure 3**. (Hint: Be sure to read the fine print in the advertisement.)
   b. Verify that, 11.12% is the effective rate that corresponds to a 10.40% nominal rate compounded daily as shown in the fine print of **Figure 3**.

8. To answer the following questions, use **Figure 4**.
   a. Verify that the nominal rate of 8% compounded daily is equivalent to the effective rate of 8.45%.
   b. Verify that you will have $1,914 at the end of 8 years using the 8% compounded daily rate.
   c. Verify that you will have $1,914 at the end of 8 years using the 8.45% effective rate.

Exercises **4** and **8** show that either the nominal rate or the effective rate can be used to determine the amount of money that you will have at the end of the investment period. Occasionally, newspaper advertisements state incorrect rates and, in turn, incorrect values of money. For example, in **Figure 5**, the effective rate is stated

Nominal vs. Effective Rates of Interest 35

BFS Bedford Federal Savings
11.15% annual percentage yield   10.40% annual interest rate
2-1/2 Year or Longer Term Account - Guaranteed*

a rate higher than any other commercial bank can pay on 2½ to 9 year term accounts opened by January 31 - minimum deposit - $500.

Today, interest rates are at an all-time high. Yet money market rates are increasingly unpredictable due to a fluctuating economy. Now is an excellent time to lock up those all-time high rates for as long as possible.

At Bedford Federal Savings you are guaranteed the high rate of interest shown above for 2½ years. This rate is higher than any commercial bank can legally pay on accounts of this type. A different high rate is offered each month for new accounts.

The risk in high yield investments is eliminated because your deposits are insured by the FDIC up to $40,000 for each account opened in a different legal capacity.

You can always withdraw the interest credited without penalty, and, if the bank consents, you can withdraw principal from the account and you forfeit only 6 month's interest on the amount withdrawn.

Act now! This rate is available only to January 31!

Interest is compounded daily, credited quarterly, and must remain on deposit one full year to earn the effective annual yield shown above. This yield reflects a 366-day year including February 29. This is a 2½-year account and the effective annual yield for the remaining 1½ years of the account is 11.12%, reflecting a normal 365-day year. This account is offered in addition to the bank's standard term accounts which are offered at varying terms and rates.

Figure 3.

You need only $1000 to earn $914

You don't need big money to earn big money at Bedford Federal Savings. You need only a minimum deposit of $1000 to earn 8%* annual interest, 8.45% annual yield on an 8 year time deposit account. This is the highest bank interest allowed by law - and $1000 is all you need to take advantage of it!

Your money grows! Put the minimum $1000 in an 8 year time deposit account and keep the principal and interest on deposit for the full term. After 8 years your savings will be worth $1,914. You earn $914 on a $1000 investment! That works out to $114 a year -- and that's growth!

*Interest on time deposit accounts is compounded daily, credited quarterly, and is available for withdrawal upon demand. If principal is withdrawn prior to maturity, federal regulations require reduction of interest on the amount withdrawn to the regular savings rate and forfeiture of 3 month's interest at that rate.

Figure 4.

11

SJM Savings

If you're serious about saving, open a 2-1/2 to 12 year savings account

THE CHOICES ARE YOURS!

If you open a 2½ to 12 year savings account you can choose the length of time you want your money to remain in your account, and you can choose the amount of your deposit, which will earn you an annual yield of 9.825% based on a rate of 9.25% compounded daily. This rate changes bi-weekly.

You choose how long after 2½ years you want to keep your money on deposit. In other words – you can choose 2½ years or more, all the way up to 12 full years.

With a $500 minimum deposit you have the flexibility to deposit any amount in excess of $500. For example, if you deposited $1000 at the current rate, after a year you would earn $112.31. With a $5000 deposit you would earn $561.67. With a $10,000 deposit you would earn $1,123.14!

Figure 5.

to be 9.825%. This is incorrect as shown by the following two methods:

Method #1: If 9.25% is the compounded daily rate, then the effective rate is

$$(1 + .0925/360)^{365} - 1 = .09831 = 9.831\%.$$

Method #2: The ad states that "if you deposited $1000 at the current rate, after just one year you would earn $112.31." Remember that the effective rate is the same as the simple interest rate for one year. (See **Exercise 2(b)**.) If the $112.31 interest amount is correct, then $I = Prt$ becomes

$$112.31 = 1000(r)(1),$$

or $r = .11231$, so that $r = 11.231\%$ is the effective rate, not 9.825%.

Exercise

9. Shortly after the newspaper advertisement in **Figure 5** appeared, some changes were made. (See **Figure 6**.) If the amounts of

SJM Savings

If you're serious about saving, open a 2-1/2 to 12 year savings account

THE CHOICES ARE YOURS!

If you open a 2½ to 12 year savings account you can choose the length of time you want your money to remain in your account, and you can choose the amount of your deposit, which will earn you an annual yield of 9.825% based on a rate of 9.25% compounded daily. This rate changes bi-weekly.

You choose how long after 2½ years you want to keep your money on deposit. In other words – you can choose 2½ years or more, all the way up to 12 full years.

With a $500 minimum deposit you have the flexibility to deposit any amount in excess of $500. For example, if you deposited $1000 at the current rate, after a year you would earn $98.31. With a $5000 deposit you would earn $491.55. With a $10,000 deposit you would earn $983.10!

Figure 6.

$98.31, $491.55 and $983.10 are correct, determine the effective rate from one of these values. Compare your result with Method 1 discussed above.

2.4 Compounding Continuously and Effective Rate

"Most banks ... compound your interest on a daily basis ... "

Most banks and savings institutions compound your interest on a daily basis. Other possible ways would be hourly, once every minute, once every second, etc. But as you will soon see, as the unit of conversion gets smaller and smaller, the amount of your investment does not increase substantially. In turn, as the number of interest periods per year gets larger and larger for a given nominal rate, the corresponding effective rates do *not* increase without bounds.

In **Example 4** we will use the formula

$$(1 + j/360n)^{360n} - 1 \qquad (F)$$

to determine the effective rate and not

$$(1 + j/360n)^{365n} - 1, \qquad (G)$$

where n is the number of *compound periods per day*. Later we will discuss the use of **(G)**.

Example 4.
Determine the effective rates if the nominal rate of 8% is compounded (a) hourly and (b) once every minute.

Solution:
  a. If 8% is compounded hourly, then the effective rate is

  $$(1 + .08/(360)(24))^{(360)(24)} - 1 = .038286 = 8.329\%.$$

  b. If the 8% is compounded once every minute, then the effective rate is

  $$(1 + .08/(360)(24)(60))^{(360)(24)(60)} - 1 = .083286$$
  $$= 8.329\%.$$

Exercise
10. If you were able to invest some money at 8% compounded every second, determine the equivalent effective rate.

Example 4 and **Exercise 10** show that as the unit of conversion gets closer to zero, the corresponding effective rates, when rounded to three places, are identical. In short, as the number of interest periods per year gets larger and larger without bound, it appears that the corresponding effective rates do *not* increase significantly. When the number of interest periods per year gets larger and larger without bound, we say that the nominal rate is *compounded continuously*.

In a calculus course, it is shown that as m becomes larger and larger without bound (this is denoted by $m \to \infty$), then $(1 + 1/m)^m$ get closer and closer to a fixed number that is denoted by e, where $e \simeq 2.718281828$. In turn,

as $m \to \infty$, then $(1 + 1/m)^m \to e$,

or, as $m \to \infty$, then $\left[(1 + 1/m)^m\right]^j \to e^j$,

or, as $m \to \infty$, then $\left(\left[(1 + 1/m)^m\right]^j - 1\right) \to (e^j - 1)$.

(H)

Let's now see how **(H)** is related to **(F)**. Returning to **(F)**, we have

effective rate $= (1 + j/360n)^{360n} - 1$,

$$= \left(1 + \frac{1}{(360/j)(n)}\right)^{(360n/j)(j)} - 1,$$

$$= \left[\left(1 + \frac{1}{(360/j)(n)}\right)^{(360/j)}\right]^{(n)j} - 1. \quad \textbf{(I)}$$

Comparing **(H)** and **(I)**, let $m = (360/j)(n)$. As $n \to \infty$, then also $m = (360/j)(n) \to \infty$, so that in **(I)**, the effective rate $\to (e^j - 1)$, or the effective rate $\to (2.718281828^j - 1)$. Hence, if the nominal rate j (expressed as a decimal) is compounded continuously, the corresponding equivalent effective rate is $(e^j - 1)$. You should remember that $(e^j - 1)$ is derived from the "360 over 360" method, not from the "365 over 360" method.

Example 5.
 a. Find the effective rate that corresponds to a nominal rate of 8% compounded continuously, using $(e^j - 1)$.
 b. Round off your result in (a) above to three decimal places and compare that effective rate with the effective rates of **Example 4** and **Exercise 10**.

Solution:
 a. Since the effective rate is given by $(e^j - 1)$ for a given compounded continuously nominal rate j expressed as a decimal, then for 8%, we have

$$(e^{.08} - 1) = (2.718281828^{.08} - 1),$$
$$= 1.083287068 - 1,$$
$$= .083287068,$$
$$= 8.3287068\%.$$

 b. Rounding to three decimal places, the effective rate is 8.329%, the same effective rate that was obtained in **Example 4** and **Exercise 10**.

Example 6.
Referring to **Figure 7**, the advertisement indicates that a 5% nominal rate compounded continuously is equivalent to 5.20% as an effective rate.

 a. Show that the stated effective rate is incorrect using $(e^j - 1)$.

 b. Indicate how the effective rate *might* be obtained had the interest been *compounded daily*.

Solution:
 a. For $5\% = .05$ compounded continuously, then the corresponding effective rate is

$$e^{.05} - 1 = (2.718281828^{.05} - 1)$$
$$= .0512710903 \quad \text{or } 5.13\%, \quad \text{not } 5.20\%.$$

Figure 7.

    b. If the 5% nominal rate is compounded daily using the "360 over 360" method, then the effective rate is

$$(1 + .05/360)^{360} - 1 = .0512674461$$

or 5.13%. If the 5% nominal rate is compounded daily using the "365 over 360" method, then the effective rate is

$$(1 + .05/360)^{365} - 1 = .0519970958$$

or

$$5.19970958\% = 5.20\%.$$

    Let's stop a moment: In developing the expression $(e^j - 1)$, we used **(F)**. What would happen had we used **(G)**? Returning to **(G)**, notice that

$$\text{effective rate} = (1 + j/360n)^{365n} - 1,$$

$$= \left(1 + \frac{1}{360n/j}\right)^{365n} - 1,$$

$$= \left(1 + \frac{1}{360n/j}\right)^{(360n/j)\cdot(365j/360)} - 1,$$

$$= \left[\left(1 + \frac{1}{360n/j}\right)^{(360n/j)}\right]^{(365j/360)} - 1. \quad \text{(J)}$$

Letting $n \to \infty$ and using the result of **(H)**, where $m = (360n/j)$ then **(J)** becomes

$$\text{effective rate} = e^{(365j/360)} - 1,$$

where j is the decimal form of the nominal rate compounded continuously; this is called the "365 over 360" method for compounding continuously.

Returning to **Example 6**, let's calculate the effective rate using this method. Then

$$\text{effective rate} = e^{(365)(.05)/360} - 1,$$

$$= 5.20013993\%,$$

or, the effective rate = 5.20%. Hence, the advertisement in **Figure 7** is correct on the basis of the "365 over 360" method, but it is incorrect if one uses the "360 over 360" method.

Notice that in the solution of **Example 6(b)**, 5% compounded daily (for a 365-day year) is equivalent to an effective rate of 5.19970958%. But 5% compounded continuously (for a 365-day year) is equivalent to an effective rate of 5.20013993%. The difference is 0.00043035%. This means that for a $1,000 investment for one year at 5% compounded continuously, you would receive only $0.43 more than when the interest is 5% compounded daily.

Exercise

11. a. Show that the effective rate of 12% compounded continuously in **Figure 8** is correct on the basis of the "365 over 360" method.

b. Compare your result for part (a) with the results of **Exercise 6**.

17

```
╔══════════════════════════════════╗
║                                  ║
║      2-1/2% year term            ║
║            certificate           ║
║                                  ║
║        MONTH OF MARCH            ║
║                                  ║
║                  effective       ║
║      12.94%      annual          ║
║                  yield           ║
║                                  ║
║                  annual          ║
║      12.00%      interest        ║
║                  rate            ║
║                                  ║
║  A full ¼% more than commercial  ║
║  bank's pay. Early withdrawal    ║
║  requires substantial penalty.   ║
║                                  ║
║   −  MINIMUM DEPOSIT $500        ║
║   −  COMPOUNDED CONTINUOUSLY     ║
║   −  RATE CHANGES MONTHLY        ║
║   −  CERTIFICATE MATURES IN 30   ║
║      MONTHS                      ║
║   −  AVAILABLE FOR INDIVIDUAL    ║
║      AND CORPORATE ACCOUNTS      ║
║                                  ║
║         HOME TOWN                ║
║          SAVINGS                 ║
╚══════════════════════════════════╝
```

Figure 8.

2.5 Nominal and Effective Rates of Inflation

Nominal and effective rates play a very important role in the discussion of inflation. You, the consumer, realize that as the rate of inflation increases, the value of your money decreases.

Inflation is measured by the Consumer Price Index (CPI). In July 1980, the CPI stood at 247.8. That means that goods and services that cost $100 in the 1967 base period cost $247.80 in July 1980. Each month, the Labor Department computes the CPI for the previous month and issues a statement as to what the percentage increase (decrease) was for the CPI from the one month to the next.

Example 7.
In **Figure 9**, is the 11.3% rate a nominal rate compounded monthly or an effective rate?

Nominal vs. Effective Rates of Interest   43

Living Costs Continuing Upward Spiral

WASHINGTON (AP) — Inflation continued at a rapid pace in May as the cost of food, housing and transportation drove consumer prices up 0.9 percent, the Labor Department said today.

The increase matched the 0.9 percent rise in April, which had been the largest jump in more than a year. In the past three months, consumer prices have gone up at a rate that would average 11.3 percent if spread out over the entire year.

Figure 9. (Courtesy of the Associated Press)

Solution:
Since 12 months per year $\times$ 0.9% per month = 10.8% per year, then 10.8% is the nominal rate compounded monthly. Also, since $(1 + 0.009)^{12} - 1 = .113509675$, or 11.3% "rounded off" (most people would round it off to 11.4%), then the 11.3% is the effective rate.

Exercise
12. In **Figure 10**, is the 12% annual increase a nominal rate compounded monthly or an effective rate?

Example 8.
Suppose that the Consumer Price Index (CPI) for a given year was 14.1% above that of the previous year. Determine what the corresponding nominal rate compounded monthly is if the 14.1% is an effective rate.

Solution:
Since $r = (1 + i)^n - 1$, where i is the rate per month, then

$$.141 = (1 + j/12)^{12} - 1,$$

19

Consumer Price Rise Causing Apprehension

> WASHINGTON (AP) — Sharp new increases in consumer prices, particularly beef and veal, are giving the Carter administration a new case of economic indigestion.
>
> Some government officials said they expected the Consumer Price Index for January, due out later today, would show a startling 1 percent increase from December prices — perhaps even more.
>
> A 1 percent monthly increase would translate to a 12 percent annual increase, but most analysts do not expect it to be that high over the course of a full year.

Figure 10. (Courtesy of the Associated Press)

where j is the nominal rate compounded monthly,

$$(1.141)^{1/12} = 1 + j/12$$
$$j = 12[(1.141)^{1/12} - 1] = .1326 = 13.3\%.$$

Exercises

13. Suppose that the effective rate of inflation is 13.3%. Express this in terms of a nominal rate compounded monthly.

14. (Optional exercise.) Each month the Department of Labor issues a statement about the percentage increase (decrease) for the CPI. Find a newspaper article that states the rate for the previous month. For that rate, determine both the nominal rate compounded monthly and the corresponding effective rate.

2.6 Nominal and Effective Rates in Other Areas

There are several other areas of our economy where the nominal and effective rates are used. One of these areas is that of wholesale prices.

Fasten Your Seat Belts As Air Fares Take Off

By JANE BRYANT QUINN

NEW YORK — Air fares today are moving so high, so fast, that it's not even possible to write about them with assurance. In surveying the outlook for 1980 vacation travel, several people didn't want to be quoted on fares for fear of being obsolete by the time this column was published.

In many cases you can nail down today's prices by buying now for a flight this summer. But some airlines reserve the right to add fuel surcharges or otherwise increase the fares, even for tickets paid in advance.

Under the law, the Civil Aeronautics Board (CAB) sets a standard industry level for U.S. fares at least once every six months. But fuel and labor costs are going up so rapidly that the CAB now computes fares every two months. And most airlines take the allowed increase.

EFFECTIVE JAN. 1, the increase permitted on U.S. fares was 3.2 percent. For March-April, the limit is 2.5 percent. If that combined rate of increase persists, average air fares this year will rise about 18 percent, compared with 32 percent in 1979 and 7 percent in 1978.

Figure 11. (Courtesy of the Buffalo Courier Express)

CHECK & SAVE!
WRITE CHECKS
EARN INTEREST

WRITE CHECKS WHILE YOUR FUNDS KEEP EARNING INTEREST RIGHT UP TO THE MINUTE THAT YOUR CHECKS CLEAR THE BANK!

EARN 5.25% ANNUAL INTEREST ON YOUR DAILY BALANCE, PAID AND COMPOUNDED MONTHLY (EFFECTIVE ANNUAL YIELD: 5.38%).

THERE IS NO SERVICE CHARGE IF YOU KEEP A $1,500 MINIMUM BALANCE (OR A $3,000 AVERAGE BALANCE).

UNITED BANK OF NEWTON **UBN**

Figure 12.

Example 9.
Suppose that for a given month wholesale prices rose 1.4%. If this was the same for each month of that year, determine the nominal rate compounded monthly and the effective rate for wholesale prices for that year.

Solution:
The nominal rate would have been $12(1.4\%) = 16.8\%$ compounded monthly. The effective rate would have been

$$(1 + .014)^{12} - 1 = .181559129, \quad \text{or} \quad 18.2\%.$$

Exercise
15. Suppose that, as an effective rate, wholesale energy goods rose 82.2% in one year. Determine the corresponding nominal rate compounded monthly.

Another area that concerns the consumer (and the manufacturer also) is that of productivity. Productivity is a measure of how many goods and services are produced in one hour of paid working time. The decline in the rate of productivity means that higher labor costs cannot be offset through increased production; this results in increased inflation.

Exercise
16. Suppose that productivity *decreased* during a given year at a rate of 1.9% compounded quarterly. Determine the corresponding effective rate of interest.

Anyone who has purchased an airplane ticket recently knows that many air fares are increasing rapidly. The Civil Aeronautics Board, which regulates air fares, now computes the fares every other month and usually reports the percentage increase on a bimonthly basis.

Exercise
17. In **Figure 11**, it is stated that air fares in the United States rose 3.2% during January and February of 1980, while for March and April, the rise was 2.5%. Verify that this yields about an 18% effective rate increase for 1980 if the combined rate of increase persists.

3. Conclusion

As you have seen in this module, the consumer must be aware of the uses (and misuses) of the terms "nominal rate" and "effective rate." These terms frequently appear in advertisements and news stories presented on television and radio and in newspapers and magazines. The next time you see or hear of a rate presented in the form of a percentage, consider how it has been used. Is it a nominal or an effective rate (or neither), has it been used correctly, and is it consistent with the other information presented?

4. Model Exam

1. Show that the effective annual yield of 5.38% in **Figure 12** is correct.

CURRENT 30 MONTH MONEY MARKET YIELDS:

FINANCIAL INSTITUTIONS	ANNUAL YIELD
First National Bank of Nashua	12.47%
Bedford Savings and Loan	12.23%
First Federal Savings of Newton	**12.94%**
American Bank of Boston	12.23%
Waltham Federal Savings Bank	12.75%
First Federal Savings of Hollis	12.75%
Western Liberty Savings & Loan	12.75%
United Bank of Burlington	12.55%

We want you to know that all money market certificates are not alike. At our bank, the effective annual yield on $100 minimum deposit, 30-month money market certificates is as high or higher than any federally insured bank or savings and loan. Guess who sponsored this add? And next time you want to invest, come to us!

Annual Yield of 12.94% based on a rate of 12.00% compounded continuously.

Rates effective through Oct. 15
Substantial penality for early withdrawal

Figure 13.

2. Consider **Figure 13** to answer each of the following:
   a. Suppose that the nominal rate is 12% compounded daily, although it is *not* stated in the advertisement. Determine which of the four methods for compounding daily is used to determine *some* of the effective rates presented in the ad.
   b. Suppose that the nominal rate is 12% compounded continuously, as stated in the advertisement. Determine which of the four methods for compounding continuously is used to determine *some* of the effective rates presented in the ad.

3. Find an error in the advertisement in **Figure 14**, if the 7% is compounded daily.

4. For a 12.2% effective rate of interest, determine the corresponding nominal rate compounded monthly.

5. Use **Figure 15** to answer each of the following:
   a. Show that the two rates presented in the ad are equivalent, if the stated annual rate is compounded daily by the "365 over 360" method and the stated annual yield is compounded semiannually.

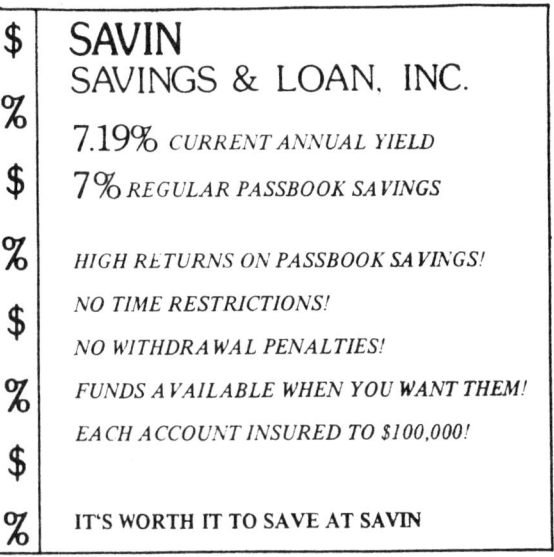

Figure 14.

12.505% annual yield   11.968% annual rate

Check the rate of your money market fund. Is the rate as high as THE EAGLE BONUS PLAN? Probably not...and, on a money market fund, your return is not guaranteed because the rate fluctuates daily. You could be losing interest.

THE EAGLE BONUS PLAN is a guaranteed investment. It's a combination $10,000 six-month certificate and separate passbook account that's guaranteed today's high rate for the next six months. And your account is insured up to $100,000 by an agency of the Federal Government, unlike money market funds!

THE EAGLE BONUS PLAN earns you the highest rate on insured savings allowed by law. The rate and yield may change at the six-month renewal date, and there is a penalty for early withdrawal.

For guaranteed high yield, invest today in THE EAGLE BONUS PLAN. You'll be glad you did!

EAGLE FEDERAL SAVINGS

Figure 15.

   b. Show that 12.505% is (almost) the effective annual yield for the 11.968% nominal rate compounded daily, using 182.5 days for half a year.

6. a. Suppose that you read the following item from a newspaper: "the effective rate of inflation for last year was 12.7% if the rate is 1% per month." Is this correct?
   b. In the same newspaper, you read that "the nominal rate of increase was 10.6% for a rate per month of 0.8%." Is this correct?

5. Answers to Exercises

1. a. $I = Prt = (\$750)(.075)(1.5) = \$84.38$

   $S = P + I = \$750 + \$84.38 = \$834.38$

   or

   $S = P(1 + rt) = \$750(1 + (.075)(1.5)) = \$834.38.$

b. $S = P(1 + i)^n = \$750(1 + .075/4)^6 = \$838.43$
$I = S - P = \$838.43 - \$750 = \$88.43.$

2. a. i. $S = \$500(1 + .105/1)^1 = \$552.50.$

ii. $S = \$500(1 + (.105)(1)) = \$552.50.$

b. Suppose that you invest P dollars for one year at an effective rate of $r\%$. After one year, then you will have

$$S = P(1 + r/1)^1$$
$$= P(1 + r)$$
$$= P(1 + rt), \text{ where } t = 1 \text{ (year)},$$

which is the same amount of money had you invested P dollars for one year at a simple interest rate of $r\%$.

3. a. If the nominal rate is 9.51% compounded weekly, then the effective rate is

$$r = (1 + .0951/52)^{52} - 1 = 9.97\%.$$

b. If the nominal rate is 9.52% compounded monthly, then the effective rate is

$$r = (1 + .0952/12)^{12} - 1 = 9.95\%.$$

c. If the nominal rate is 9.53% compounded quarterly, then the effective rate is

$$r = (1 + .0953/4)^4 - 1 = 9.88\%.$$

Hence, (a) is the best.

4. a. $S = P(1 + i)^n = \$1000(1 + .0814/1)^3 = \$1264.62.$
In **Example 2**, $S = \$1264.57$; the two values differ only because 8.14% has been rounded to two places.

b. $S = P(1 + i)^n = \$1000(1 + .0806/1)^3 = \$1261.81.$
In **Example 2**, $S = \$1261.67$; the two values differ only because 8.06% has been rounded to two places.

5. For the PAL Savings and Loan, use a daily rate of $j\%/360$ and a year with 360 days.

8.00% daily: effective rate $= (1 + .08/360)^{360} - 1 = 8.33\%$

7.50% daily: effective rate $= (1 + .075/360)^{360} - 1 = 7.79\%$

10.40% daily: effective rate $= (1 + .104/360)^{360} - 1 = 10.96\%$

6.50% daily: effective rate $= (1 + .065/360)^{360} - 1 = 6.72\%$

6.00% daily: effective rate $= (1 + .06/360)^{360} - 1 = 6.18\%$

5.50% daily: effective rate $= (1 + .055/360)^{360} - 1 = 5.65\%.$

Nominal vs. Effective Rates of Interest   51

For the Lion Federal Savings, use a daily rate of $i\%/360$ and a year with 365 days.

7.50% daily: $(1 + .075/360)^{365} - 1 = 7.90\%$

6.75% daily: $(1 + .0675/360)^{365} - 1 = 7.08\%$

6.50% daily: $(1 + .065/360)^{365} - 1 = 6.81\%$.

6. **Case 1**: $(1 + .12/360)^{360} - 1 = 12.747\%$

   Case 2: $(1 + .12/365)^{365} - 1 = 12.747\%$

   Case 3: $(1 - .12/360)^{365} - 1 = 12.94\%$

   Case 4: $(1 + .12/365)^{360} - 1 = 12.56\%$.

   As in **Example 3**, Cases 1 and 3 are illustrated in **Figure 2**. In Case 1, a 360 day year is used, while in Case 3, a 365 day year is used. In both cases, the rate is 12%/360 per day.

7. **a.** effective rate = $(1 + .104/360)^{366} - 1 = 11.15\%$. Here a 366 day year is used.
   b. effective rate = $(1 + .104/360)^{365} - 1 = 11.12\%$. Here a 365 day year is used.

8. **a.** effective rate = $(1 + .08/360)^{365} - 1 = 8.45\%$.
   b. $S = P(1 + i)^n = \$1000(1 + .08/360)^{2922} = \$1914.13$, or $1914. (Note: 2922 days = $8 \times 365 + 2$.)
   c. $S = P(1 + i)^n = \$1000(1 + .0845/1)^8 = \$1913.54$, or $1914.

9. **$98.31**: $I = Prt$ becomes $\$98.31 = (\$1000)(r)(1)$, or $r = 9.831\%$.

   $491.55: $I = Prt$ becomes $\$491.55 = (\$500)(r)(1)$, or $r = 9.831\%$.

   $983.10: $I = Prt$ becomes $\$983.10 = (\$10,000)(r)(1)$, or $r = 9.831\%$.

   Each of these effective rates is the same as that obtained in Method 1.

10. effective rate = $(1 + .08/(360)(24)(60)(60))^{(360)(24)(60)(60)} - 1$

    $= 8.329\%$.

11. **a.** If 12% is compounded continuously for a 365 day year, then the effective rate = $e^{(365)(.12)/(360)} - 1 = 12.94\%$.
    b. To two decimal places, 12% compounded continuously for a 365 day year is equivalent to 12% compounded daily for a 365 day year.

12. A 1% monthly increase means that the nominal rate is $(1\%)(12) = (12\%)$ compounded monthly, or this is a 12% annual increase. The effective rate = $(1 + .01)^{12} - 1 = 12.68\%$.

13. Since $r = (1 + i)^n - 1$, then $1 + .133 = (1 + j/12)^{12}$, where j is the nominal rate compounded monthly,

$$(1.133)^{1/12} = 1 + j/12,$$
$$j/12 = (1.133)^{1/12} - 1,$$
$$j = 12[(1.133)^{1/12} - 1],$$
$$j = .1255$$

or, $j = 12.55\%$ compounded monthly.

14. (Optional exercise.)

15. Since $r = (1 + i)^n - 1$, then $1 + .822 = (1 + j/12)^{12}$, where j is the nominal rate compounded monthly,

$$(1.822)^{1/12} = 1 + j/12,$$
$$j/12 = (1.822)^{1/12} - 1,$$
$$j = 12[(1.822)^{1/12} - 1],$$
$$j = 61.52\% \text{ compounded monthly.}$$

16. effective rate $= (1 + (-.019)/4)^4 - 1,$
$$= -1.89\%.$$

Hence, the effective rate is a decrease of 1.89%.

17. effective rate $= [(1 + .032)(1 + .025)]^3 - 1,$
$$= 18.36\%.$$

6. Answers to Model Exam

1. effective rate $r = (1 + .0525/12)^{12} - 1 = 5.38\%.$

2. a. "360 over 360" daily: $r = (1 + .12/360)^{360} - 1 = 12.75\%$

"360 over 365" daily: $r = (1 + .12/365)^{360} - 1 = 12.56\%$

"365 over 360" daily: $r = (1 + .12/360)^{365} - 1 = 12.94\%$

"365 over 365" daily: $r = (1 + .12/365)^{365} - 1 = 12.75\%.$

b. "360 over 360" continuously: $r = e^{(360)(.12)/(360)} - 1$
$$= 12.75\%$$

"360 over 365" continuously: $r = e^{(360)(.12)/(365)} - 1$
$$= 12.56\%$$

"365 over 360" continuously: $r = e^{(365)(.12)/(360)} - 1$
$$= 12.94\%$$

"365 over 365" continuously: $r = e^{(365)(.12)/(365)} - 1$
$$= 12.75\%.$$

3. If 7% is compounded daily, then the four possible effective rates are

$$r = (1 + .07/360)^{360} - 1 = 7.25\%$$
$$(1 + .07/365)^{360} - 1 = 7.14\%$$
$$(1 + .07/360)^{365} - 1 = 7.35\%$$
$$(1 + .07/365)^{365} - 1 = 7.25\%$$

Note: If the compounded daily rate is 6.85%, then the corresponding effective rate by the "365 over 360" method is

$$r = (1 + .0685/360)^{365} - 1 = 7.19\%.$$

4. If j is the nominal rate compounded monthly, then

$$(1 + .122) = (1 + j/12)^{12}$$
$$(1.122)^{(1/12)} = (1 + j/12)$$
$$j = 12((1.122)^{(1/12)} - 1)$$
$$j = 11.57\%.$$

5. **a.** $(1 + .11968/360)^{365} - 1 = 12.90\%$, and

$$(1 + .12505/2)^2 - 1 = 12.90\%.$$

b. $2((1 + .11968/360)^{182.5} - 1) = 12.508\%.$

6. **a.** effective rate $r = (1 + .01)^{12} - 1 = 12.68\%$, or 12.7%. Hence, the stated effective rate is correct.

b. $12 \times (0.8\%) = 9.6\%$. Hence, the stated nominal rate is incorrect.

UMAP Module 518

Modules in Undergraduate Mathematics and its Applications

Oligopolistic Competition

Donald R. Sherbert

Published in cooperation with the Society for Industrial and Applied Mathematics, the Mathematical Association of America, the National Council of Teachers of Mathematics, the American Mathematical Association of Two-Year Colleges, The Institute of Management Sciences, and the American Statistical Association.

INTERMODULAR DESCRIPTION SHEET:	UMAP Unit 518
TITLE:	OLIGOPOLISTIC COMPETITION
AUTHOR:	Donald R. Sherbert Dept. of Mathematics University of Illinois Urbana, IL 61801
MATHEMATICAL FIELD:	Intermediate calculus
APPLICATION FIELD:	Economics
TARGET AUDIENCE:	Students in an intermediate calculus or mathematical modeling course.
ABSTRACT:	The word monopoly comes from the Greek stem polein, meaning to sell, with the prefix "mono-," indicating one. Thus, it is used to describe a market structure in which there is only one seller of a good. The term oligopoly refers to a market in which a small number of firms compete. A change in price or production level by one firm will cause reaction in the others. The particular case of only two competitors is called a duopoly. A number of mathematical models have been developed to analyze the competitive structure of oligopolies. In this unit we examine some of the elementary models for duopolies that utilize calculus.
PREREQUISITES:	Computing partial derivatives; determination of maximum values of functions of several variables.

© Copyright 1982, 1987 by COMAP, Inc. All rights reserved.

COMAP, 60 Lowell Street, Arlington, MA 02174 (617) 641-2600

Oligopolistic Competition

Donald R. Sherbert
Department of Mathematics
University of Illinois
Urbana, IL 61801

Table of Contents

1. Introduction 1
2. The Cournot Model 2
3. The Chamberlin Model 4
4. Duopolies, Reaction Curves 5
5. The Leader-Follower Model 7
6. Sample Test 9
7. Solutions to Exercises 9
8. Solutions to Sample Test 11
    References 12

Modules and Monographs in Undergraduate Mathematics and Its Applications (UMAP) Project

The goal of UMAP was to develop, through a community of users and developers, a system of instructional modules in undergraduate mathematics and its applications to be used to supplement existing courses and from which complete courses may eventually be built.

The Project was guided by a National Advisory Board of mathematicians, scientists, and educators. UMAP was funded by a grant from the National Science Foundation and is now supported by the Consortium for Mathematics and Its Applications (COMAP), Inc., a nonprofit corporation engaged in research and development in mathematics education.

COMAP Staff

Paul J. Campbell	Editor
Solomon A. Garfunkel	Executive Director, COMAP
Laurie W. Aragon	Business Development Manager
Philip A. McGaw	Production Manager
Theresa Cronin	Copy Editor
Annemarie S. Morgan	Administrative Assistant
John Gately	Distribution
Hannah Vincent	Secretary

UMAP Advisory Board

Steven J. Brams	New York University
Llayron Clarkson	Texas Southern University
Donald A. Larson	SUNY at Buffalo
R. Duncan Luce	Harvard University
Frederick Mosteller	Harvard University
George M. Miller	Nassau Community College
Walter Sears	University of Michigan Press
Arnold A. Strassenburg	SUNY at Stony Brook
Alfred B. Willcox	Mathematical Association of America

This module was developed under the auspices of the UMAP Economic Panel, chaired by Gerald Egerer of Sonoma State University. The Project would like to thank W.T. Fishback of Earlham College, M. Eisenberg of the University of Massachusetts at Amherst, Robert E. Girse of Idaho State University, and Erwin Eltze of Fort Hays State University for their reviews, and all others who assisted in the production of this unit.

This material was prepared with the partial support of National Science Foundation (NSF) Grant No. SED76-19615 A02 and No. SED80-07731. It is based on work supported by the National Science Foundation under CAUSE Grant No. SER78-06381 to the University of Illinois at Urbana–Champaign. Recommendations expressed are those of the author and do not necessarily reflect the views of the NSF or the copyright holder.

1. Introduction

The word *monopoly* comes from the Greek stem polein, meaning to sell, with the prefix "mono-" indicating one. Thus, it is used to describe a market structure in which there is only one seller of a good. The only monopolies in the United States, such as utilities companies, operate under government regulations and restrictions.

"... oligopoly refers to a market in which a small number of firms compete."

The term *oligopoly* refers to a market in which a small number of firms compete. The prefix "oligo-" means few, so that an oligopoly is "competition among the few." An oligopolistic market is one in which the behavior of one firm will have an impact on the behavior of the competing firms. A change in price or production level by one firm will cause a reaction in the others. The particular case of only two competitors is called a *duopoly*.

Many markets in the United States are oligopolistic in nature. Oligopolies can be separated into two broad categories. In one type, such as the markets for aluminum, steel, petroleum, and heavy electrical equipment, the product is fairly *homogeneous* and main purchasers are other industries. The other type, referred to as a *differential oligopoly*, attempts to distinguish similar products by brand names, styling, advertising, and so on. Examples of such oligopolies are the markets for automobiles, liquor, cigarettes, television sets, and toothpaste. In these cases, the market is dominated by a half dozen or fewer firms and their decisions determine the entire market. Even though some smaller firms may exist that produce a nominally competitive good, they follow the leaders and have negligible impact on the market as a whole.

A number of mathematical models have been developed to analyze the competitive structure of oligopolies. Unfortunately, none of them portrays general practices with great accuracy, especially for differential oligopolies. This does not mean these models are worthless; any model attempting to describe a complicated social or economic phenomenon focuses on particular aspects and hinges on basic assumptions. A model can focus on certain patterns, allowing various assumptions to be pursued and their logical consequences analyzed.

In this module we examine some of the elementary models for duopolies which utilize calculus. For a more complete discussion, including models based on the theory of games, see the references given at the end of this module. The mathematical tools needed for this unit are the calculation of partial derivatives and the determination of maximum values of functions of several variables.

2. The Cournot Model

The first attempt to analyze the nature of duopoly was by the French economist A. Cournot in 1838. His pioneering work did not attract much attention at the time; but when it was translated into English and published in 1897, the economists of that period responded with critical interest. Cournot did not express his ideas in terms of calculus, but such a formulation is easily done and we do so here.

To study the case of two competitors, Cournot imagined two neighboring mineral springs whose owners sold the water to people who brought their own containers. Thus there were no costs of production to consider. He also assumed a linear demand curve

$$p = a - bx \quad (a, b > 0),$$

where x is the total number of gallons sold each day and p is the price per gallon. Let x_1, x_2 denote the respective owner's sales per day, so that $x = x_1 + x_2$. The problem is to determine the equilibrium output for each owner and the resulting price. The main point of interest is the comparison of these results with the case of a monopoly.

"The problem is to determine the equilibrium output for each owner and the resulting price."

Let us consider a monopoly first. If only one mineral spring is in operation, then the owner's revenue each day is

$$R = xp = x(a - bx).$$

To maximize R, we set the derivative dR/dx of R (known as marginal revenue) equal to 0. Thus

$$\frac{dR}{dx} = a - 2bx = 0,$$

so that the optimal level for a monopoly is

$$x_M = \frac{a}{2b}.$$

It is easily seen that this value of x does actually maximize R since the second derivative is negative. The resulting price is

$$P_M = \frac{a}{2}.$$

Now suppose the second mineral spring opens for business. The ultimate sales levels and prices will presumably maximize the revenue of each owner. If x_1, x_2 denote the sales at each spring per day at price $p = a - b(x_1 + x_2)$, then the revenues of each are given respectively by

$$R_1 = x_1 p$$
$$= x_1[a - b(x_1 + x_2)]$$

and

$$R_2 = x_2 p$$
$$= x_2[a - b(x_1 + x_2)].$$

At this stage we shall make a behavioral assumption known as the *Cournot hypothesis*: Each owner believes that the other owner will not change his output, even though each may observe changes by the other. In other words, the first owner will maximize R_1 with respect to x_1 while regarding x_2 as constant. The second owner behaves similarly. Proceeding under this assumption to see what consequences will follow, we have

> *"Each owner believes that the other owner will not change his output...."*

$$\frac{\partial R_1}{\partial x_1} = a - 2bx_1 - bx_2 = 0$$

and

$$\frac{\partial R_2}{\partial x_2} = a - bx_1 - 2bx_2 = 0.$$

Solving these equations simultaneously, we obtain the optimal levels

$$x_1 = x_2 = \frac{a}{3b},$$

which are readily verified as maximizing R_1 and R_2. The corresponding price is then

$$p = \frac{a}{3}.$$

Comparing these results to those for a monopoly, $x_M = a/2b$ and $p_M = a/2$, we see that the total output for the duopoly, $x = x_1 + x_2 = 2a/3b$, is larger than that for a monopoly, and the price $p = a/3$ is smaller. One could conclude from this analysis that the consumer benefits more from a duopoly than a monopoly.

Exercise 1

Suppose the demand function for a good is $p = f(x)$, where $f'(x) \neq 0$. Suppose duopolists produce x_1, x_2 respectively, where $x = x_1 + x_2$. Using the Cournot assumption, show that if the revenues $R_1 = x_1 f(x_1 + x_2)$ and $R_2 = x_2 f(x_1 + x_2)$ are maximized, then $x_1 = x_2$.

Exercise 2

Suppose there are n competing mineral springs where $n \geq 3$, and let $x_1, x_2, \ldots, x_n$ denote the sales per day. Assuming the demand curve is linear as in the preceding discussion, and using the Cournot behavioral assumption, find the values of $x_1, x_2, \ldots, x_n$ that maximize revenues and find the resulting price. What happens as n gets larger?

3. The Chamberlin Model

A classic study of oligopolies is Edward Chamberlin's *The Theory of Monopolistic Competitions* [Chamberlin 1933]. It was one of the first in-depth analyses of competition among the few. Part of the book is devoted to an examination of the Cournot model. Chamberlin suggests that the owners will recognize their market interdependence and will respond in a more sophisticated way.

"... the owners will recognize their market interdependence...."

Using the results obtained from the Cournot model, we see that each duopolist would get optimal revenue equal to

$$R = x_1 p = \frac{a}{3b} \cdot \frac{a}{3} = \frac{1}{9}\left[\frac{a^2}{b}\right].$$

In the case of a monopoly, the optimal revenue at $x_M = a/2b$ is

$$R_M = x_M p_M = \frac{a}{2b} \cdot \frac{a}{2} = \frac{1}{4}\left[\frac{a^2}{b}\right].$$

A smart duopolist immediately sees that half of the monopoly revenue, which is

$$\frac{1}{2} R_M = \frac{1}{8}\left[\frac{a^2}{b}\right],$$

is larger than the revenue predicted by the Cournot solution for a duopoly. Thus it would be more profitable for the duopolists to cooperate and share the monopoly revenue. This is called *collusion*.

This observation often reflects reality more closely than Cournot's conclusion. In practice, oligopolies tend to have a spirit of cooperation rather than a fiercely competitive attitude.

Exercise 3
Compare the revenues for an oligopoly of n firms with monopoly revenue. Is it better to share the monopoly revenue? (See **exercise 2**.)

4. Duopolies, Reaction Curves

We now extend the study of duopolies to include cost of production. It is assumed that the market for a particular commodity has two sellers, firm A and firm B. If firm A produces x_1 units and firm B produces x_2 units, then the total output of the two firms is $x = x_1 + x_2$ units. It is assumed that there is a relationship between market price p and output x given by a demand function

$$p = f(x) = f(x_1 + x_2).$$

Normally, price will decrease as supply is increased, so it is assumed that p is a decreasing function of x. The price p is determined by the total output x, but each firm's output affects $x = x_1 + x_2$, and so has an impact on the price.

The total revenue generated by selling x units at a price p per unit is xp, quantity times price. The revenue coming into each firm is given by

Firm A: $R_1 = x_1 p = x_1 f(x_1 + x_2),$
Firm B: $R_2 = x_2 p = x_2 f(x_1 + x_2).$

However, each firm also incurs production costs that depend on the level of output. Let $C_1(x_1)$ represent the total cost for firm A to produce output x_1, and let $C_2(x_2)$ be the cost function for firm B.

"The profit enjoyed by a firm is its incoming revenue minus its costs."

The profit enjoyed by a firm is its income revenue minus its costs. Thus the profits Π_1 and Π_2 for the firms are given by

Firm A: $\Pi_1 = x_1 f(x_1 + x_1) - C_1(x_1),$
Firm B: $\Pi_2 = x_2 f(x_1 + x_2) - C_2(x_2).$

It is assumed that each firm wishes to establish a production level that will maximize its own profits. The complicating factor is that each time one firm changes its output, the other firm will react. This action and reaction will affect total output, which affects the market price, which affects revenue, which affects profits.

Suppose firm B is producing at level x_2. Firm A, viewing x_2 as fixed, wishes to find x_1 such that

$$\frac{\partial \Pi_1}{\partial x_1} = f(x_1 + x_2) + x_1 f'(x_1 + x_2) - C_1'(x_1) = 0.$$

This equation defines x_1 as a function of x_2: Given x_2, we let $x_1 = x_1^*$ be the solution of $\partial \Pi_1/\partial x_1 = 0$. The resulting function $x_1 = \alpha(x_2)$ is the *reaction curve* for firm A. It tells firm A how to adjust to any change in firm B.

Similarly, if firm B views x_1 as given, then the equation

$$\frac{\partial \Pi_2}{\partial x_2} = f(x_1 + x_2) + x_2 f'(x_1 + x_2) - C_2'(x_2) = 0$$

determines a reaction curve $x_2 = \beta(x_1)$ for firm B.

Suppose that the graphs of the reaction curves $x_1 = \alpha(x_2)$ and $x_2 = \beta(x_1)$ intersect in a point (x_1^*, x_2^*) as illustrated in **Figure 1**. At this point each firm is behaving optimally with respect to the other, and consequently no change will occur. Such a point is called a *Cournot equilibrium*.

If we take costs to be zero, so that $C_1(x_1) = C_2(x_2) = 0$, and let $f(x) = a - bx$, then the resulting reaction curves are found to be

$$x_1 = \alpha(x_2) = \frac{a}{2b} - \frac{1}{2}x_2,$$

$$x_2 = \beta(x_1) = \frac{a}{2b} - \frac{1}{2}x_1.$$

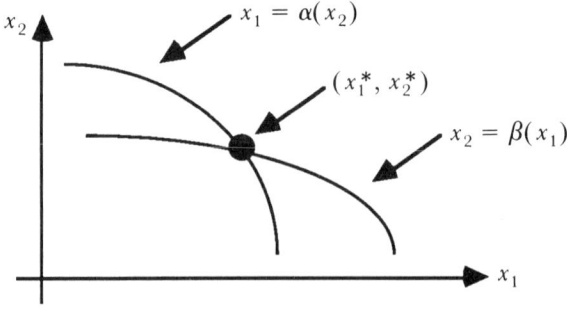

Figure 1

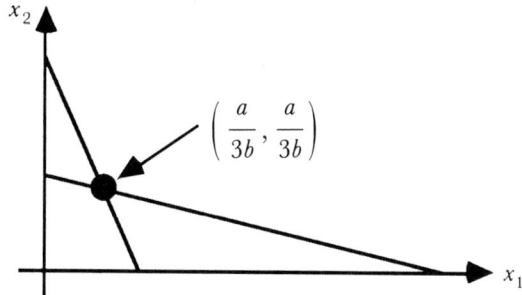

Figure 2

Solving these equations simultaneously, we find that the Cournot equilibrium is $x_1^* = x_2^* = a/3b$, which is precisely the Cournot solution found earlier. (See **Figure 2**.)

Exercise 4
Assume a demand function $p = f(x) = 12 - x$ (where $x = x_1 + x_2$) and cost functions

$$C_1(x_1) = \tfrac{1}{2}x_1^2$$

for firm A and

$$C_2(x_2) = \tfrac{1}{2}x_2^2$$

for firm B. Find the reaction curves for each firm, and find the Cournot equilibrium. Also find the total production level x and price p.

Exercise 5
Collusion occurs when duopolists act as a two-plant monopolist and maximize $\Pi = \Pi_1 + \Pi_2$. Using the demand and cost functions of **exercise 4**, find the collusion solution by solving $\partial \Pi / \partial x_1 = 0$, $\partial \Pi / \partial x_2 = 0$. Compare to the result of **exercise 4** relative to production level and price.

5. The Leader-Follower Model

The approach proposed by Heinrich von Stackelberg is based on the assumption that one firm of a duopoly is the leader and the other firm is the follower. The leader sets its production level and assumes that the follower will respond according to its reaction curve. The leader ignores its own reaction curve, but it uses the follower's reaction curve to maximize profits.

Suppose firm A is the leader. Its profit function is a function of both x_1 and x_2, say

$$\Pi_1 = G(x_1, x_2).$$

Since firm A assumes the reaction of firm B to be governed by its reaction curve $x_2 = \beta(x_1)$, it can regard its profit function to be a function of x_1 alone, namely,

$$\Pi_1 = G(x_1, \beta(x_1)).$$

Maximizing this function of one variable then determines the optimal production level x_1^* for firm A. Firm B then produces at level $x_2^* = \beta(x_1^*)$.

In the case that $p = f(x) = b - ax$ and costs are 0, we found that firm A's profit function was

$$\Pi_1 = x_1(a - b(x_1 + x_2))$$

and firm B's reaction was

$$x_2 = \frac{a}{2b} - \frac{1}{2}x_1.$$

Substitution of the expressions for x_2 into Π_1 gives us

$$\Pi_1 = \tfrac{1}{2}ax_1 - \tfrac{1}{2}bx_1^2.$$

Then

$$\frac{d\Pi_1}{dx_3} = \frac{1}{2}a - bx_1 = 0,$$

so that the Stackelberg solution is $x_1 = a/2b$. Firm B reacts with $x_2 = a/4b$.

In this case $p = a/4$, and the revenue for firm A is $a^2/8b$, which is the same revenue as in collusion. But note that the revenue for firm B is $a^2/16b$, which is less than collusion. Firm A benefits from being the leader and firm B loses.

Exercise 6
Using the functions of **exercise 4**, and assuming firm A to be the leader, find the Stackelberg solution for maximum profits.

6. Sample Test

Suppose a duopoly has demand function
$$p = f(x_1 + x_2) = 200 - (x_1 + x_2)$$
and cost functions $C_1(x_1) = 10x_1$ for firm A, $C_2(x_2) = x_2^2$ for firm B.

1. Find the point of Cournot equilibrium.

2. Find the collusion solution.

3. Find the leader-follower solution assuming firm A to be the leader.

7. Solutions to Exercises

1. From
$$\frac{\partial R_1}{\partial x_1} = f(x_1 + x_2) + x_1 f'(x_1 + x_2) = 0,$$

$$\frac{\partial R_2}{\partial x_2} = f(x_1 + x_2) + x_2 f'(x_1 + x_2) = 0$$

we get
$$x_1 f'(x_1 + x_2) = x_2 f'(x_1 + x_2),$$

and therefore, $x_1 = x_2$. (We assume the demand curve does not have zero slope.)

2. For $k = 1, 2, \ldots, n$, we have $R_k = x_k(a - b(x_1 + \cdots + x_n))$. Then for each k, we have

$$\frac{\partial R_k}{\partial x_k} = a - b(x_1 + \cdots + x_n) - bx_k = 0.$$

Subtracting equation i from equation j gives us $x_i = x_j$ for $i \neq j$. Adding the n equations gives us

$$na = (n+1)b(x_1 + \cdots + x_n) = n(n+1)bx_k,$$

and thus we get

$$x_k = \frac{1}{n+1} \cdot \frac{a}{b}$$

for each k. Also $p = a/(n+1)$. These approach 0 as n tends to ∞.

3. For n firms in an oligopoly, the Cournot model gives each firm

$$R = \left(\frac{1}{n+1} \cdot \frac{a}{b}\right)\left(\frac{1}{n+1}a\right) = \frac{1}{(n+1)^2}\frac{a^2}{b}.$$

Monopoly sharing gives each firm

$$\frac{1}{n}R_M = n\left(\frac{1}{n}\frac{a}{b}\right)\left(\frac{a}{n}\right) = \frac{1}{n^3}\frac{a^2}{b}.$$

If $n \geq 3$, then $(1/n)R_M$ is less than R.

4. Since

$$\pi_1 = x_1(12 - x_1 - x_2) - \left(\frac{1}{2}\right)x_1^2,$$

we have

$$\frac{\partial \Pi_1}{\partial x_1} = 12 - 3x_1 - x_2 = 0,$$

so that

$$x_1 = \frac{1}{3}(12 - x_2).$$

Similarly, since

$$\pi_2 = x_2(12 - x_1 - x_2) - \frac{1}{3}x_2^2,$$

we have

$$\frac{\partial \Pi_2}{\partial x_2} = 12 - x_1 - \frac{8}{3}x_2,$$

so that

$$x_2 = \frac{3}{8}(12 - x_1).$$

Equilibrium occurs at $x_1 = 20/7 = 2.86$, $x_2 = 24/7 = 3.4$, with price $p = 40/7 = 5.7$.

5. Since

$$\pi = (x_1 + x_2)(12 - x_1 - x_2) - \frac{1}{2}x_1^2 - \frac{1}{3}x_2^2,$$

we get

$$\frac{\partial \pi}{\partial x_1} = 12 - 3x_1 - 2x_2 = 0,$$

$$\frac{\partial \pi}{\partial x_2} = 12 - 2x_1 - \frac{8}{3}x_2 = 0,$$

and thus $x_1 = 2$, $x_2 = 3$ with price $p = 7$. Production levels are smaller and price larger than in **exercise 4**.

6. From **exercise 4** we have

$$x_2 = \tfrac{3}{8}(12 - x_1),$$

so that

$$\pi_1 = x_1(12 - x_1 - x_2) - \tfrac{1}{2}x_1^2$$
$$= x_1\left(\tfrac{15}{2} - \tfrac{5}{8}x_1\right) - \tfrac{1}{2}x_1^2.$$

We then get

$$\frac{d\pi_1}{dx_1} = \frac{15}{2} - \frac{9}{2}x_1 = 0,$$

so that $x_1 = 5/3$. Then $x_2 = 31/8$.

8. Solutions to Sample Test

1. Since

$$\pi_1 = x_1(200 - x_1 - x_2) - 10x_1,$$
$$\pi_2 = x_2(200 - x_1 - x_2) - x_2^2,$$

we have

$$\frac{\partial \pi_1}{\partial x_1} = 190 - 2x_1 - x_2 = 0,$$

$$\frac{\partial \pi_2}{\partial x_2} = 200 - x - 4x_2 = 0.$$

We find that the Cournot equilibrium occurs at $x_1 = 93.5$, $x_2 = 3$.

2. To maximize total profit

$$\pi = (x_1 + x_1)(200 - x_1 - x_2) - 10x_1 - x_2^2,$$

we solve

$$\frac{\partial \pi}{\partial x_1} = 190 - 2x_1 - 2x_2 = 0,$$

$$\frac{\partial \pi}{\partial x_2} = 200 - 2x_1 - 4x_2 = 0$$

and find that $x_1 = 90$, $x_2 = 5$.

11

3. Putting $x_2 = (1/4)(200 - x_1)$ into π gives us

$$\pi_1 = x_1\left(150 - \tfrac{3}{4}x_1\right) - 10x_1,$$

so that

$$\frac{d\pi_1}{dx_1} = 140 - \tfrac{3}{2}x_1 = 0.$$

Hence $x_1 = 280/3 = 93.3$, and then $x_2 = 80$.

References

Cournot, A. 1897. *Researches into the Mathematical Principles of the Theory of Wealth*.

Chamberlin, E.H. 1933. *The Theory of Monopolistic Competition*.

Ferguson, C. *Microeconomics*.

Henderson, J., and R. Quandt. 1971. *Microeconomic Theory, a Mathematical Approach*, 2nd ed.

Kogiku, K.C. 1971. *Macroeconomic Models*.

Varian, R. 1984. *Microeconomics Analysis*, 2nd ed. New York: W. W. Norton.

UMAP Module 526

Modules in Undergraduate Mathematics and its Applications

Dimensional Analysis

Frank R. Giordano,
Michael E. Wells,
Carroll O. Wilde

Published in cooperation with the Society for Industrial and Applied Mathematics, the Mathematical Association of America, the National Council of Teachers of Mathematics, the American Mathematical Association of Two-Year Colleges, The Institute of Management Sciences, and the American Statistical Association.

COMAP

INTERMODULAR DESCRIPTION SHEET:	UMAP Unit 526
TITLE:	Dimensional Analysis
AUTHORS::	Frank R. Giordano Dept. of Mathematics U.S. Military Academy West Point, NY 10996 Michael E. Wells 13055 Farthingale Drive Herndon, VA 22071-2622 Carroll O. Wilde Dept. of Mathematics Naval Postgraduate School Monterey, CA 93940
REVIEW STAGE / DATE:	IV 5/13/83
CLASSIFICATION:	Applications in Linear Algebra, Math Modeling and Engineering
PREREQUISITE SKILLS:	1. Ability to solve systems of linear algebraic equations. 2. Working knowledge of basic linear algebra concepts such as rank of a matrix and linear independence. 3. Rudimentary knowledge of basic physical quantities, such as velocity, acceleration, kinetic energy.
OUTPUT SKILLS:	Given a physical system and the factors that influence it: 1. Be able to find a complete set of dimensionless products for the system; 2. Apply Buckingham's theorem and determine a dimensionally correct equation relating the variables involved in the system.
RELATED UNITS:	Keeping Dimensions Straight (Unit 564)

© Copyright 1983, 1988 COMAP, Inc. All rights reserved.

COMAP, 60 Lowell Street, Arlington, MA 02174 (617) 641-2600

Dimensional Analysis

Frank R. Giordano
Department of Mathematics
U.S. Military Academy
West Point, NY 10996

Michael E. Wells
13055 Farthingale Drive
Herndon, VA 22071-2622

Carroll O. Wilde
Department of Mathematics
Naval Postgraduate School
Monterey, CA 93940

Table of Contents

1. INTRODUCTION 1
2. DIMENSIONS AS PRODUCTS 2
3. FORMING DIMENSIONLESS PRODUCTS 5
4. MAXIMAL SETS 7
5. COMPLETE SETS 11
6. BUCKINGHAM'S THEORY 12
7. AN ALGORITHM 14
8. CONCLUSION 18
9. REFERENCES 18
10. MODEL EXAM 19
11. ANSWERS TO EXERCISES 19
12. ANSWERS TO MODEL EXAM 22

MODULES AND MONOGRAPHS IN UNDERGRADUATE
MATHEMATICS AND ITS APPLICATIONS (UMAP) PROJECT

The goal of UMAP was to develop, through a community of users and developers, a system of instructional modules in undergraduate mathematics and its applications to be used to supplement existing courses and from which complete courses may eventually be built.

The Project was guided by a National Advisory Board of mathematicians, scientists, and educators. UMAP was funded by a grant from the National Science Foundation and is now supported by the Consortium for Mathematics and Its Applications (COMAP), Inc., a non-profit corporation engaged in research and development in mathematics education.

COMAP STAFF

Paul J. Campbell	Editor
Solomon A. Garfunkel	Executive Director, COMAP
Laurie W. Aragon	Business Development Manager
Philip A. McGaw	Production Manager
Theresa Cronin	Copy Editor
Annemarie S. Morgan	Administrative Assistant
John Gately	Distribution

The Project would like to thank Michael J. Kallaher of Washington State University, Jon Laible of Eastern Illinois University, and Walter Meyer of Adelphi University for their reviews, and all others who assisted in the production of this unit.

This unit was field-tested and/or student-reviewed in preliminary form by Stanley C. Leja of the U.S. Military Academy at West Point, Professor Snider of the University of South Florida, Derek Wong of Northern Illinois University, Frank Giordano of the Naval Postgraduate School, Richard Alo' of Lamar University, Michael Penna of Indiana-Purdue University-Indianapolis, R. McFadden of Northern Illinois University, and Peter Nicholls of Northern Illinois University, and was revised on the basis of data received from these sites.

This material was prepared with the partial support of National Science Foundation Grant No. SED76-19615 A02, SED80-07731, and No. SPE-8304192. Recommendations expressed are those of the author and do not necessarily reflect the views of the NSF or the copyright holder.

1. Introduction

"The central problem is to establish a relationship...."

In mathematical modeling we are concerned with relationships that describe properties and behavior of systems that arise in physical, social, or other sciences. For a given system, we are often interested in some characteristic that is represented by a certain variable, called the *dependent* variable, and we identify other variables, called the *independent* variables, that affect the dependent variable. The central problem is to establish a relationship between the dependent variable and the independent variables which describes properties and behavior of the system.

As an example, suppose that the given physical system is the undamped, unforced simple pendulum shown in **Figure 1**. One characteristic that is vital in understanding the behavior of a pendulum is its period, which we represent by the dependent variable t. We can readily identify as possible independent variables the length of the pendulum l, its mass m, the initial angular displacement from the vertical α, and the acceleration due to gravity g. The basic problem is to establish a relationship between the dependent variable t and the independent variables l, m, α and g. (We shall study this example further in Section 8.)

The pendulum example displays a scenario that is typical of mathematical modeling. Once the dependent variable has been specified, the modeler begins by identifying all possible candidates for independent variables, using his or her knowledge of physical principles and other relevant background or experience. It is crucial to produce a list that contains all the essential variables, but the initial list may contain superfluous items as well. Once we have a candidate set of variables, *dimensional analysis* becomes a powerful technique to help the modeler establish the relationship among the variables.

Dimensional analysis is based on the following principles: that physical quantities have dimensions, and that physical laws are unaltered when the units for measuring the dimensions are changed. For example, the area A of a rectangle is the product of its base b by its height h, whether the unit of length be metric, English or other; i.e., the law $A = bh$ holds regardless of the unit of length used to measure b and h. These two principles provide the basis for a technique that yields the following outcome;

> The result of a dimensional analysis is an equation that relates the dependent variable to some or all of the independent variables and that is valid for any system of units in which the variables are measured.

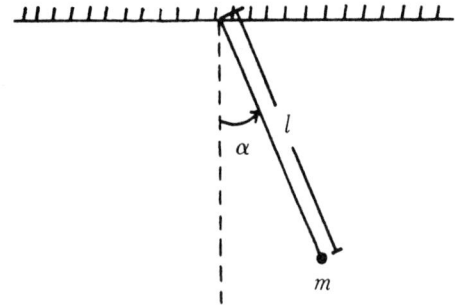

Figure 1. A simple pendulum.

"The result of a dimensional analysis is an equation...."

Often the equation that results directly from the dimensional analysis is not the final form that is desired, and additional experimental work may be necessary. We do hope, however, that the result of a dimensional analysis will serve as a basis for the design of appropriate experiments to determine the correct relationship.

2. Dimensions as Products

The three primary physical quantities we consider are mass, length and time, and with these quantities we associate dimensions M, L and T, respectively. Some other quantities are secondary in the sense that they are usually defined directly in terms of mass, length and/or time; their dimensions are determined by the algebraic operations involved in the definitions. Thus, since velocity may be defined as the ratio of distance (dimension L) traveled to time (dimension T) of travel, i.e., $v = lt^{-1}$, the dimension of velocity is LT^{-1}. Similarly, since area is fundamentally a product of length (dimension L) by width (also of dimension L), its dimension is L^2.

There are still other quantities that are more complex in the sense that they are not usually defined directly in terms of mass, length and time alone, but instead their definitions include other quantities, such as velocity. We associate dimensions with these more complex quantities again in accordance with the algebraic operations involved in the definitions. For example, since momentum is the product of mass by velocity, its dimension is $M(LT^{-1})$, or simply MLT^{-1}.

The basic definition of quantity may also involve dimensionless constants; these are ignored in finding dimensions. Thus, the dimension of kinetic energy, which is one half (a dimensionless constant) the product of mass by velocity squared, has dimension $M(LT^{-1})^2$ or, simply, ML^2T^{-2}.

Exercise
1. Find the dimension of:
   a. acceleration, a ratio of velocity to time;
   b. linear mass density, a ratio of mass to distance;
   c. volume mass density, a ratio of mass to volume.

These examples illustrate some important concepts regarding dimensions of physical quantities.

(a) There are three physical quantities upon which the concept of dimension is based: mass m, length l, and time t. These quantities are measured in some appropriate system of units whose choice does not affect the assignment of dimensions.

(b) There are other physical quantities, such as area and velocity, that are defined as simple products involving only mass, length and/or time. Here we use the term "product" to include any quotient, since we may indicate division by negative exponents.

(c) There are still other, more complex, physical quantities, such as momentum and kinetic energy, whose definitions involve quantities other than mass, length and time. But since the simpler quantities from (a) and (b) are themselves products, these more complex quantities can also be expressed as products involving mass, length and time by algebraic simplification. We shall use the term *product* to refer to any physical quantity from item (a), (b) or (c); a product from (a) is trivial, since it has only one factor.

(d) To each product is assigned a *dimension*, i.e., an expression of the form

$$M^m L^p T^q, \qquad (1)$$

where n, p and q are real numbers that may be positive, negative or zero.

When a basic dimension is missing from a product, the corresponding exponent is understood to be zero. Thus, the dimension $M^2 L^0 T^{-1}$ may also appear as $M^2 T^{-1}$. When n, p and q are all zero in an expression of the form (1), so the dimension reduces to

$$M^0 L^0 T^0, \qquad (2)$$

the quantity, or product, is said to be *dimensionless*.

Exercise

2. For a sector of a circle (see **Figure 2**) with central angle θ and radius r, the length of arc s subtended by θ is given by $s = r\theta$. Using this formula, show that the trivial product L^0 results when we try to find the dimension of θ (i.e., θ is a dimensionless quantity).

Recall from Section 1 that the basic result of a dimensional analysis should be an equation that relates the variables of the system under study. Such equations normally include expressions that are obtained by performing mathematical operations on products of variables. Thus, in general, these equations may involve sums, square roots, logarithms and trigonometric functions of products, for example.

Special care must be taken in forming sums of products because, just as we cannot "add apples and oranges," in an equation we cannot add products that have unlike dimensions. For example, if f denotes force, m mass and v velocity, we know immediately that the equation

$$f = mv + v^2$$

cannot be correct because mv has dimension MLT^{-1} while v^2 has dimension L^2T^{-2}. These dimensions are unlike, and hence the products mv and v^2 cannot be added. An equation such as this, i.e., one that contains among its terms two products having unlike dimensions, are said to be *dimensionally incompatible*. Equations that involve only sums of products having the same dimension are, then, dimensionally compatible.

The concept of dimensional compatibility is related to another important concept, called *dimensional homogeneity*. In general, an equation that is true regardless of the system of units in which the variables are measured is said to be *dimensionally homogeneous*. In particular, if an equation involves only sums of dimensionless products (i.e., it is a dimensionless equation), then the equation is dimensionally homogeneous. For, since the products are dimensionless, the

> "...we cannot add products that have unlike dimensions."

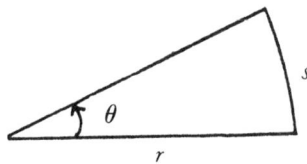

Figure 2. Sector of a circle.

factors used for conversion from one system of units to another would simply cancel. For equations that involve only sums of products, dimensional homogeneity is the special case of dimensional compatibility in which each product is dimensionless.

In dimensional analysis we determine the form of a desired equation basically by finding an appropriate dimensionless equation and then solving for the dependent variable. More precisely, for a system under study we determine the relationship among the variables by a procedure that includes the following tasks:

(i) determine all dimensionless products among the variables (in general there may be infinitely many such products, so they must be described rather than actually written out);

(ii) use the dimensionless products to find all possible dimensionally homogeneous equations among the variables;

(iii) use the dimensionally homogeneous equations to obtain the final relationship.

We shall present examples to illustrate methods of carrying out these tasks in the remaining sections of this module.

Exercise

3. Determine whether the equation

$$s = s_0 + v_0 t - \tfrac{1}{2} g t^2$$

is dimensionally compatible, if s is the position (measured vertically from a fixed reference point) of a body at time t, s_0 the position at $t = 0$, v_0 the initial velocity and g the acceleration due to gravity.

3. Forming Dimensionless Products

To illustrate a method of forming all dimensionless products from the variables in a given system, we use an example from [Noble 1967]. This example illustrates task (i) from Section 2 only; no attempt is made to carry the larger modeling problem to a complete solution.

Example 1.

Suppose we have a smooth spherical body that is submerged in a fluid. Specific illustrations include water droplets in air, air bubbles

Figure 3. Fluid flow around a submerged spherical body of diameter δ.

in water, fine sediment in air or water, and even weather balloons in the air. In general, the fluid flow about such an object (see **Figure 3**) depends upon the motion of the fluid, the motion of the body, or a combination of the two.

For simplicity, we consider the case in which only the fluid is moving. Then the following variables are relevant: the fluid velocity v, its density ρ, its viscosity μ, the acceleration due to gravity g, and the diameter of the body δ. The dimensions of these variables are given in **Table 1**.

We wish to find all possible dimensionless products among these variables. Since any product of these variables is of the form

$$v^a \rho^b \delta^c g^d \mu^e, \tag{3}$$

where a, b, c, d, e are real numbers, we have from **Table 1** and the discussion in Section 2 that any such product has the form

$$(LT^{-1})^a (ML^{-3})^b (L)^c (LT^{-2})^d (ML^{-1}T^{-1})^e. \tag{4}$$

Simplifying the expression (4) we obtain

$$M^{b+e} L^{a-3b+c+d-e} T^{-a-2d-e}. \tag{5}$$

Since a dimensionless product must also have the form $M^0 L^0 T^0$ (see (2) above), we may equate exponents to obtain the system of equations

$$\begin{cases} \quad\quad\;\; b \quad\quad\quad\quad\; + e = 0 \\ a - 3b + c + \;\, d - e = 0 \\ -a \quad\quad\quad\; - 2d - e = 0. \end{cases} \tag{6}$$

Table 1.
Dimensions of the fluid-flow variables

Variable	v	ρ	δ	g	μ
Dimension	LT^{-1}	ML^{-3}	L	LT^{-2}	$ML^{-1}T^{-1}$

Any solution of the system of equations **(6)** yields a dimensionless product. For example, one solution is given by ($a = b = c = 1$, $d = 0$, $e = -1$); and substitution of these values into **(3)** yields the dimensionless product $v\rho\delta/\mu$, which is known as the *Reynolds number*. Another solution of the system **(6)** that yields a dimensionless product important enough in fluid dynamics to have a special name given by ($a = 2$, $b = 0$, $c = d = -1$, $e = 0$). The corresponding dimensionless product is $v^2/\delta g$, and its positive square root is called the *Froude number*.

In the next section, we continue the study of dimensionless products, applying some basic results of linear algebra to the problem. We close this section with an exercise, following a brief remark on the significance of the Reynolds and Froude numbers.

In the motion of a viscous fluid about a submerged object, the Reynolds number can be interpreted as the ratio of the inertia forces, as represented by the product ρv^2 (twice the kinetic energy of unit volume), to the viscous forces $\mu v/\delta$, which is the product of the coefficient of viscosity by a velocity gradient. The Froude number F is defined by $F^2 = v^2/\delta g$, which may be interpreted as the ratio of the inertia forces to the mechanical forces.

Exercises

4. For **Example 1** find two dimensionless products other than the Reynolds and Froude numbers.

5. Find a dimensionless product relating the torque $\tau(ML^2T^{-2})$ produced by an automobile engine, the engine's rotation rate $\omega(T^{-1})$, the volume of air displaced by the engine $v(L^3)$, and the air density $\rho(ML^{-3})$.

4. Maximal Sets

"... we seek all possible dimensionless products ...."

Example 1 shows that in general many dimensionless products may be formed from the variables of a given system. Each such product corresponds to a solution of the homogeneous system of linear algebraic equations obtained by equating the exponents of M, L and T in the simplified dimension expression to zero. Since we seek all possible dimensionless products, we would like to find a linearly independent set of solutions of the system from which we can obtain all other solutions by forming linear combinations; such a set is called a *maximal linearly independent set* of solutions. We shall abbreviate this name, calling such a set simply a *maximal set*.

We note first that the system (6) is a special case of a homogeneous linear system

$$A\vec{x} = 0,$$

where A is an $m \times n$ matrix and $\vec{x}$ is an $n \times 1$ (column) vector. From linear algebra we know that the number of solutions in a maximal set equals $n - r$, where r is the rank of A. Applying this result to our 3×5 system (6), we first determine the rank of the coefficient matrix by row reduction:

$$\begin{bmatrix} 0 & 1 & 0 & 0 & 1 \\ 1 & -3 & 1 & 1 & -1 \\ -1 & 0 & 0 & -2 & -1 \end{bmatrix} \sim \begin{bmatrix} 1 & 0 & 0 & 2 & 1 \\ 0 & 1 & 0 & 0 & 1 \\ 0 & 0 & 1 & -1 & 1 \end{bmatrix}.$$

Thus, the rank is 3 and there are $5 - 3 = 2$ solutions in a maximal set.

For example, the solutions $[(a = b = c = 1, d = 0, e = -1)$ and $(a = -2, b = 0, c = d = 1, e = 0)]$ are linearly independent as we can see from the values of d and e.

Since a maximal set has two solutions, we may choose as arbitrary variables any two of the unknowns for which the 3×3 matrix obtained from the coefficient matrix by deleting the columns corresponding to these unknowns is nonsingular, and then solve for the three remaining variables in terms of them. Examination of the coefficient matrix in our example reveals that b and e cannot both be chosen arbitrarily since removal of the second and fifth columns would leave a singular matrix. Variables d and e, on the other hand, may be chosen arbitrarily. The row reduction used to determine the rank of the coefficient matrix also yields the system

$$\begin{cases} a & + 2d + e = 0 \\ & b + e = 0 \\ & c - d + e = 0, \end{cases}$$

from which we find a, b, and c in terms of d and e:

$$\begin{cases} a = -2d - e \\ b = - e \\ c = d - e. \end{cases} \tag{7}$$

For the particularly convenient choices $(d = 1, e = 0)$ and $(d = 0, e = 1)$ we obtain the solutions

$$(a = -2, b = 0, c = d = 1, e = 0), \tag{8}$$

and

$$(a = b = c = -1, d = 0, e = 1), \tag{9}$$

respectively, from (7).

The solution (8) yields the dimensionless product

$$\delta g v^{-2} = \frac{\delta q}{v^2}, \tag{10}$$

and the solution (9) yields the dimensionless product

$$v^{-1}\rho^{-1}\delta^{-1}\mu = \frac{\mu}{v\rho\delta}. \tag{11}$$

We now consider linear combinations of the basic solutions (8) and (9). For example, by adding (8) and (9) we obtain the solution

$$(a = -3, b = -1, c = 0, d = e = 1), \tag{12}$$

which yields the dimensionless product

$$v^{-3}\rho^{-1}g\mu = \frac{g\mu}{v^3\rho}. \tag{13}$$

But the product (13) can also be obtained directly from (10) and (11) by multiplication:

$$\left(\frac{\delta q}{v^2}\right) \times \left(\frac{\mu}{v\rho\delta}\right) = \left(\frac{g\mu}{v^3\rho}\right). \tag{14}$$

Thus, addition of the solutions (8) and (9) produces a solution that yields the same dimensionless product as does multiplication of the corresponding products (10) and (11). The reason for this result is that when we add solutions for the unknowns a, b, c, d, and e, we are ultimately adding exponents (in the product (3)); and addition of exponents in algebra corresponds to multiplication of numbers with the same base: $x^m x^n = x^{m+n}$.

In forming linear combinations of solutions, we may also multiply them by scalars. For example, if we multiply the solution (9) by -1, we obtain the solution

$$(a = b = c = 1, d = 0, e = -1), \tag{15}$$

which corresponds to the dimensionless product

$$v\rho\delta\mu^{-1} = \frac{v\rho\delta}{\mu}. \tag{16}$$

This product can also be obtained directly from the product (11) by raising it to the power -1:

$$\left(v^{-1}\rho^{-1}\delta^{-1}\mu\right)^{-1} = v\rho\delta\mu^{-1}. \tag{17}$$

Thus, multiplication of the solution (9) by -1 produces a solution that yields the same dimensionless product as does raising the corresponding product (11) to the power -1. The reason for this result is that when we multiply a solution for the unknowns a, b, c, d, and e by a scalar, we are ultimately multiplying exponents (in (3)) by that scalar; and in ordinary algebra, multiplication of exponents by a scalar corresponds to raising a power to a power $(x^m)^n = x^{mn}$.

Exercise
6. For the same system of equations (6), find the dimensionless products corresponding to:
   a. the solution obtained by multiplying the solution (9) by 2;
   b. the solution obtained by subtracting (8) from (9);
   c. The sum of $1/2$ times the solution (8) and 2 times the solution (9).

For ease of notation, we regard the solutions (8) and (9) as column vectors, and denote them by β_1 and β_2, respectively. Thus, we have

$$\beta_1 = \begin{bmatrix} -2 \\ 0 \\ 1 \\ 1 \\ 0 \end{bmatrix}, \quad \beta_2 = \begin{bmatrix} -1 \\ -1 \\ -1 \\ 0 \\ 1 \end{bmatrix}.$$

It is extremely important to understand clearly the effect of forming a linear combination of β_1 and β_2.

The dimensionless products corresponding to β_1 and β_2 are $v^{-2}g\delta$ and $v^{-1}\rho^{-1}\delta^{-1}\mu$, respectively, and the dimensionless product corresponding to the solution $k_1\beta_1 + k_2\beta_2$ is

$$\left(v^{-2}g\delta\right)^{k_1}\left(v^{-1}\rho^{-1}\delta^{-1}\mu\right)^{k_2}.$$

Therefore, *addition of solutions results in multiplication of their corresponding products, and multiplication of a solution by a scalar results in raising the corresponding product to the power given by that scalar.*

Exercise

7. a. For the same system of equations **(6)**, find a second linearly independent set of solutions.
   b. Describe the effect of forming linear combinations of these solutions.

"For those of you who are familiar with the concept of a vector space...."

For those of you who are familiar with the concept of a basis of a vector space, let us examine what we have done in these terms. The solutions of the system **(6)** form a vector space of dimension 2. This vector space of solutions is a two-dimensional subspace of the vector space $V_5(R)$ of 5-tuples of real numbers. If we can find a basis for this subspace, then any other solution can be expressed as a linear combination of the basis vectors. By choosing ($d = 1$ and $e = 0$) and $d = 0$ and $e = 1$) and solving for a, b and c in each case, we determine a basis B for the solution space:

$$B = \left\{ \beta_1 = \begin{bmatrix} -2 \\ 0 \\ 1 \\ 1 \\ 0 \end{bmatrix}, \beta_2 = \begin{bmatrix} -1 \\ -1 \\ -1 \\ 0 \\ 1 \end{bmatrix} \right\}.$$

Any solutions can then be expressed in the form $k_1\beta_1 + k_2\beta_2$, where k_1, k_2 are real numbers.

5. Complete Sets

Our basic goal thus far has been to find all possible dimensionless products among the variables of a given system. The algebraic system of equations which yields these products usually has an infinite number of solutions, but we can obtain all solutions by forming linear combinations of a maximal set.

The set of dimensionless products corresponding to a maximal set is called a *complete set*. Since all linear combinations of solutions to the (homogeneous) system of equations involved are also solutions and they represent exponents of the original variables, it follows that all dimensionless products can be obtained also by forming all possible products and powers of the members of a complete set.

Exercises

8. a. Verify that the products $\delta g v^{-2}$ and $v^{-1}\rho^{-1}\delta^{-1}\mu$ form a complete set for **Example 1**, and find a second complete set.

b. Is $\{v^{-1}\rho^{-1}\delta^{-1}\mu, v\rho\delta\mu^{-1}\}$ a complete set for **Example 1**? Justify your answer.

9. The volume flow rate q for laminar flow in a pipe depends on the pipe radius $r(L)$, the fluid viscosity $\mu(ML^{-1}T^{-1})$ and the pressure drop per unit length

$$\frac{d\rho}{dz}\left(\frac{ML^{-1}T^{-2}}{L}\right).$$

Show that any complete set consists of a single dimensionless product, and that

$$\pi_1 = \frac{q}{\dfrac{d\rho}{dz}\dfrac{1}{\mu}r^4}$$

is one dimensionless product for this problem.

6. Buckingham's Theorem

In Section 2 we defined an equation to be dimensionally homogeneous if it remains true regardless of the system of units in which the variables are measured. The basic result in dimensional analysis which allows us to construct dimensionally homogeneous equations from complete sets of dimensionless products is *Buckingham's theorem*, which is stated in [Bender 1978] as follows (a proof is given in [Bridgman 1931]):

An equation is dimensionally homogeneous if and only if it can be put into the form

$$f(\pi_1, \pi_2, \ldots, \pi_n) = 0, \qquad (19)$$

where n is a positive integer, f is a function of n variables, and $\pi = \{\pi_1, \pi_2, \ldots, \pi_n\}$ is a complete set of dimensionless products.

For the case in which a complete set consists of a single dimensionless product, i.e., $\pi = \{\pi_1\}$, (19) reduces to the form

$$f(\pi_1) = 0.$$

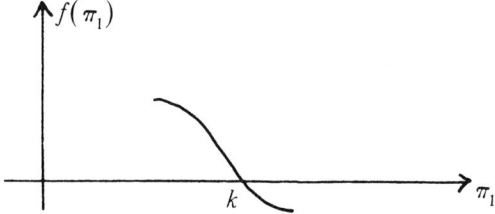

Figure 4. Graphical representation of a function $f(\pi_1)$ with one real zero, at $\pi_1 = k$.

In this case we assume that the function f has one real zero at k, as illustrated in **Figure 4** (to assume otherwise has little physical meaning). We then obtain the solution $\pi_1 = k$.

For example, in **Exercise 9** we obtained the complete set

$$\left\{ \frac{q}{\frac{d\rho}{dz}\frac{1}{\mu}r^4} \right\}.$$

Thus, for this example the equation $\pi_1 = k$ becomes

$$\frac{q}{\frac{d\rho}{dz}\frac{1}{\mu}r^4} = k,$$

from which

$$q = k\frac{1}{\mu}\frac{d\rho}{dz}r^4$$

where k is dimensionless.

For the case $n = 2$ in Buckingham's Theorem, **(19)** takes the form

$$f(\pi_1, \pi_2) = 0. \tag{20}$$

"...the dependent variable will appear in only one...."

As we shall see from the technique used in **Example 2**, we can generally choose the products in a complete set so that the dependent variable will appear in only one of them, say π_2. We then proceed under the assumption that **(20)** can be solved for the chosen product, π_2, in terms of the remaining product π_1. Such a solution takes the form

$$\pi_2 = h(\pi_1), \tag{21}$$

and once h is known we can solve **(21)** for the dependent variable. There is, of course, the problem of determining the function h. In Section 7 we shall see how experiments may be designed to determine the nature of this function.

7. An Algorithm

In this section we organize various ideas presented so far into a methodology for discovering relationships among the variables of a given system. First let us recall the main ideas that we need for the algorithm.

Candidate Set: the set of all factors (variables and dimensional constants) that influence the physical system in question.

Complete Set: a set of dimensionless products of the variables in the candidate set, from which all other dimensionless products can be obtained by considering the powers and products of these dimensionless products.

Buckingham's Theorem: an equation is dimensionally homogeneous if and only if it can be put into the form $f(\pi_1, \pi_2, \ldots, \pi_n) = 0$, where f is some function and $\pi = \{\pi_1, \pi_2, \ldots, \pi_n\}$ is a complete set of dimensionless products.

To discover the relationship between the dependent and independent variables, we may apply the following algorithm, which represents a slight reorganization of tasks (i)–(iii) listed in Section 2.

> 1. Determine a candidate set of factors (variables and constants) that influence the system under study.
>
> 2. Determine a complete set of dimensionless products $\pi = \{\pi_1, \pi_2, \ldots, \pi_n\}$.
>
> 3. Find an equation of the form $f(\pi_1, \pi_2, \ldots, \pi_n) = 0$, which exists by Buckingham's Theorem.
>
> 4. Solve this equation for the dependent variable.

Example 2.
We return to the pendulum problem of Section 1, presenting two solutions based on this algorithm. (See **Figure 1** for details.)

Step 1. The first step of the algorithm was carried out in Section 1; the candidate set S is given by

$$S = \{m, g, t, l, \alpha\}.$$

Step 2. Next we find a complete set of dimensionless products. We begin this task by listing the dimensions of the relevant variables in a table.

Variable	m	g	t	l	α
Dimension	M	LT^{-2}	T	L	dimensionless

Any product of these variables must be of the form

$$m^a g^b t^c l^d \alpha^e \tag{22}$$

and hence must have dimension

$$(M)^a (LT^{-2})^b (T)^c (L)^d.$$

Therefore, a product of the form (22) is dimensionless if and only if we have

$$M^a L^{b+d} T^{c-2b} = M^0 L^0 T^0. \tag{23}$$

Thus, we need to solve the system of equations represented by the matrix equation

$$\begin{bmatrix} 1 & 0 & 0 & 0 & 0 \\ 0 & 1 & 0 & 1 & 0 \\ 0 & -2 & 1 & 0 & 0 \end{bmatrix} \begin{bmatrix} a \\ b \\ c \\ d \\ e \end{bmatrix} = \begin{bmatrix} 0 \\ 0 \\ 0 \end{bmatrix}. \tag{24}$$

Since the rank of the coefficient matrix in (24) is 3, we seek a maximal set that consists of two solutions. The matrix formed from the coefficient matrix by deleting the second and fifth columns is particularly convenient and has a non-zero determinant. Choosing ($b = 0$, $e = 1$) and ($b = 1$, $e = 0$), we obtain the maximal set

$$B = \left\{ \begin{bmatrix} 0 \\ 0 \\ 0 \\ 0 \\ 1 \end{bmatrix}, \begin{bmatrix} 0 \\ 1 \\ 2 \\ -1 \\ 0 \end{bmatrix} \right\}.$$

Now substitute these solutions into (22) to obtain a complete set π of dimensionless products:

$$\pi_1 = m^0 g^0 t^0 l^0 \alpha^1 = \alpha, \qquad \pi_2 = m^0 g^1 t^2 l^{-1} \alpha^0 = \frac{gt^2}{l};$$

$$\pi = \{\pi_1, \pi_2\} = \left\{\alpha, \frac{gt^2}{l}\right\}. \tag{25}$$

Step 3. We now apply Buckingham's Theorem, which asserts that the variables are related by an equation of the form $f(\pi_1, \pi_2) = 0$:

$$f\left(\alpha, \frac{gt^2}{l}\right) = 0. \tag{26}$$

Step 4. According to the discussion in Section 6 (see Equation (21)), we can solve (26) to obtain an expression of the form $\pi_2 = h(\pi_1)$:

$$\frac{gt^2}{l} = h(\alpha). \tag{27}$$

Solving (27) for the period t, we obtain

$$t = \sqrt{\frac{l}{g} h(\alpha)} = k(\alpha)\sqrt{\frac{l}{g}}, \tag{28}$$

where $k(\alpha) = \sqrt{h(\alpha)}$.

Note that (28) represents only a general form for the relationship among the variables m, g, t, l and α. However, we can conclude from this expression that t does not depend on m and is related to $l^{1/2}$ and $g^{-1/2}$ by some function of α. Knowing this much, we can determine the nature of the function $k(\alpha)$ experimentally. For example, (28) shows that by keeping the ratio l/g constant, we have the relation

$$\frac{t_1}{t_2} = \frac{k(\alpha_1)}{k(\alpha_2)}$$

for any two sets of corresponding values of t and α. Then by measuring t experimentally for various values of α, we can determine the *nature* of $h(\alpha)$. For any given pendulum, l/g is constant, and for a small angular displacement α we have $k(\alpha) \simeq 2\pi$

(a constant function), from which

$$t \simeq 2\pi\sqrt{\frac{l}{g}}.$$

> "...that ultimately we wish to solve for the dependent variable...."

We note that in **Example 2** the system of equations represented by **(24)** was so easy that we solved it essentially by inspection. In more-complicated situations, it may be appropriate to use other methods from linear algebra. For example, the Gauss-Jordan method provides a systematic procedure when a solution is not apparent by inspection. In applying other methods, however, remember that ultimately we wish to solve for the dependent variable; and therefore we want it to appear in only one of the dimensionless products in a complete set. This result can be achieved by taking the dependent variable as one of our arbitrary variables and assigning convenient values to all the arbitrary variables.

Similarly, computational advantages can be achieved by choosing as arbitrary variables those that are not part of the identity submatrix of the reduced form. This choice ensures that the submatrix formed by deleting the columns corresponding to the arbitrary variables is non-singular. Furthermore, if for each arbitrary variable we assign the value one to it and the value zero to the remaining arbitrary variables we obtain a linear independent set of solutions. In our previous example, since the period is the dependent variable and we wish it to be arbitrary, we list it last. That is, we formally alter the result of Step 1 to read

$$S = \{m, g, l, \alpha, t\}.$$

Now row-reduce the augmented matrix of the corresponding system of linear equations for the exponents:

$$\begin{bmatrix} 1 & 0 & 0 & 0 & 0 & | & 0 \\ 0 & 1 & 1 & 0 & 0 & | & 0 \\ 0 & -2 & 0 & 0 & 1 & | & 0 \end{bmatrix} \underset{\sim}{\text{row}} \begin{bmatrix} 1 & 0 & 0 & 0 & 0 & | & 0 \\ 0 & 1 & 0 & 0 & -\frac{1}{2} & | & 0 \\ 0 & 0 & 1 & 0 & \frac{1}{2} & | & 0 \end{bmatrix},$$

from which we obtain the maximal set

$$\left\{ \begin{bmatrix} 0 \\ 0 \\ 0 \\ 1 \\ 0 \end{bmatrix}, \begin{bmatrix} 0 \\ \frac{1}{2} \\ -\frac{1}{2} \\ 0 \\ 1 \end{bmatrix} \right\}.$$

These solutions yield as a complete set

$$\pi = \{\alpha, g^{1/2}l^{-1/2}t\}.$$

We now continue on to Steps 3 and 4 as in **Example 2**.

Exercise

10. This problem involves a modification of the pendulum model to account for friction in the system. Suppose that the friction force f is proportional to the square of the velocity v, i.e., $f = kv^2$, where k is a constant of proportionality.

  a. Let K be the dimension of the constant k. Set up a dimensional equation from the relation $f = kv^2$, and solve for K. (Note that by Newton's second law the dimension of the force f must be $M(LT^{-2})$.)

  b. Let t be the time required for the pendulum to move from its initial position with angular displacement α to an angular displacement of $\alpha/2$. (Consider t as a modified period.) Show that the variables are related by an equation of the form

$$t = \sqrt{\frac{l}{g}}\, h\!\left(\alpha, \frac{kl}{m}\right).$$

8. Conclusion

"...the efficiency of the experimental work can be improved through dimensional analysis."

Example 2 illustrates a characteristic feature of dimensional analysis. Normally the modeler who is studying a given system has an intuitive idea of the variables involved and has a working knowledge of general principles and laws (such as Newton's second law), but lacks the precise laws governing the interaction of the variables. Of course, the modeler can always experiment with each independent variable separately, holding the others constant and measuring the effect on the system. Often, however, the efficiency of the experimental work can be improved through an application of dimensional analysis. For a more detailed discussion of the relationship between dimensional analysis and experimental design see [Langhaar 1951]. For other further study of dimensional analysis, see the references.

9. References

Bender, Edward A. 1978. *An Introduction to Mathematical Modeling.* New York: Wiley.

Birkoff, Garrett. *Hydrodynamics, A Study in Logic, Fact and Similitude.* New York: Dover.

Bridgman, Percy Williams. 1931. *Dimensional Analysis*. New Haven and London: Yale University Press.

Langhaar, Henry L. 1951. *Dimensional Analysis and Theory of Models*. New York: Wiley.

Noble, Ben. 1967. *Applications of Undergraduate Mathematics in Engineering*. New York: Macmillan.

10. Model Exam

1. The lift force ϕ (MLT^{-2}) on a missile depends upon its length l (L) velocity v (LT^{-1}), diameter δ (L), angle of attack α (dimensionless), the density ρ (ML^{-3}), viscosity μ ($ML^{-1}T^{-1}$), and speed of sound a (LT^{-1}) of the air. Show that

$$\phi = \rho v^2 l^2 h\left(\frac{\delta}{l}, \alpha, \frac{\mu}{g v l}, \frac{a}{v}\right).$$

11. Answers to Exercises

1. a. LT^{-2}; b. ML^{-1}; c. ML^{-3}.

2. $\theta = s/r = sr^{-1}$, and since s and r both have dimension L, θ has dimension LL^{-1}, or L^0.

3. The given equation is dimensionally compatible, since the dimension of each product involved is L: s and s_0 have dimension L; $v_0 t$ has dimension $(LT^{-1})T = L$; $(1/2)gt^2$ has dimension $(LT^{-2})T^2 = L$.

4. There are infinitely many possible answers to this exercise. One pair of solutions for the exponents is $[(a = b = c = 2, d = 0, e = -2), (a = -4, b = 0, c = d = 2, e = 0)]$. These solutions yield the dimensionless products

$$\frac{v^2 \rho^2 \delta^2}{\mu^2}, \quad \frac{\delta^2 g^2}{v^4}, \quad \text{respectively.}$$

5. First list the dimensions of the system variables in a table.

Variable	ω	v	ρ	τ
Dimension	T^{-1}	L^3	ML^{-3}	ML^2T^{-2}

Any product $\omega^a v^b \rho^c \tau^d$ of these variables has dimension

$$\left(T^{-1}\right)^a \left(L^3\right)^b \left(ML^{-3}\right)^c \left(ML^2T^{-2}\right)^d = M^{c+d}L^{3b-3c+2d}T^{-a-2d}.$$

Set the exponents on M, and L and T equal to zero to obtain the system of equations

$$\begin{cases} c + d = 0 \\ 3b - 3c + 2d = 0 \\ -a \qquad\quad - 2d = 0. \end{cases}$$

From these equations we see that $c = -d$, $b = -(5/3)d$ and $a = -2d$. We may assign a value to d arbitrarily, and the value $d = 3$ is convenient since it removes the fraction in b. With this value we obtain the solution ($a = -6$, $b = -5$, $c = -3$, $d = 3$), which yields the dimensionless product

$$\frac{\tau^3}{\omega^6 v^5 \rho^3}.$$

6. a. $\dfrac{\mu^2}{v^2\rho^2\delta^2}$;   b. $\dfrac{v\mu}{\rho\delta^2 g}$;   c. $\dfrac{\mu^2 g^{1/2}}{v^3\rho^2\delta^{3/2}}$.

7. a. There are many possible choices: one particular maximal set is given by

$$B = \{B_1, B_2\} = \left\{ \begin{bmatrix} -4 \\ 0 \\ 2 \\ 2 \\ 0 \end{bmatrix}, \begin{bmatrix} -2 \\ -2 \\ -2 \\ 0 \\ 2 \end{bmatrix} \right\}.$$

b. A linear combination of the solutions, $k_1 B_1 + k_2 B_2$, corresponds to the product

$$\left(\frac{\delta^2 g^2}{v^4}\right)^{k_1} \left(\frac{\mu^2}{v^2\rho^2\delta^2}\right)^{k_2},$$

which is dimensionless for any choice of k_1, k_2.

8. a. In Section 7 we found that a maximal set for the system of (6) has two elements. Since the two solutions $(-2, 0, 1, 1, 0)$ and $(-1, -1, -1, 0, 1)$ form a linearly independent set, they must form a maximal set, and hence the corresponding products $\delta g v^{-2}$ and $v^{-1}\rho^{-1}\delta^{-1}\mu$ form a complete set for this system.

There are many possible answers to this exercise; for example, the solutions indicated in the answer to **Exercise 7** yield the complete set

$$\left\{ \frac{\delta^2 g^2}{v^4}, \frac{\mu^2}{v^2\rho^2\delta^2} \right\}.$$

b. No. The dimensionless products $v^{-1}\rho^{-1}\delta^{-1}\mu$ and $v\rho\delta\mu^{-1}$ correspond to the solutions $(-1, -1, -1, 0, 1)$ and $(1, 1, 1, 0, -1)$, which are linearly dependent.

9.

Variable	r	μ	$\dfrac{d\rho}{dz}$	q
Dimension	L	$ML^{-1}T^{-1}$	$ML^{-2}T^{-2}$	$L^3 T^{-1}$

The dimension of a product

$$r^a \mu^b \left(\frac{d\rho}{dz}\right)^c q^d$$

is

$$(L)^a (ML^{-1}T^{-1})^b (ML^{-2}T^{-2})^c (L^3 T^{-1})^d = M^{b+c} L^{a-b-2c+3d} T^{-b-2c-d}.$$

Set the exponents equal to zero to obtain the system

$$\begin{cases} b + c & = 0 \\ a - b - 2c + 3d = 0 \\ -b - 2c - d = 0. \end{cases}$$

Since the coefficient matrix has rank 3, there will be $4 - 3 = 1$ solution in a maximal set, and hence one product in a complete set of dimensionless products. The solution $(a = -4, b = 1, c = -1, d = 1)$ yields the product

$$\pi_1 = \frac{q}{\dfrac{d\rho}{dz} \dfrac{1}{\mu} r^4}.$$

10. a. From the equation $f = kv^2$, we form the dimensional equation

$$MLT^{-2} = KL^2 T^{-2},$$

from which $K = ML^{-1}$.

b. Step 1. The set of relevant variables is given by $S = \{\alpha, m, l, g, k, t\}$.
Step 2. First, list the dimensions of the variables in a table

α	m	l	g	k	t
—	M	L	LT^{-2}	ML^{-1}	T

Any product $\alpha^a m^b l^c g^d k^e t^f$ has dimension $M^{b+e} L^{c+d-e} T^{-2d+f}$, hence we obtain the system of equations

$$\begin{cases} b & + e & = 0 \\ c + d - e & = 0 \\ -2d & + f = 0. \end{cases}$$

We choose a, c and f as the arbitrary variables for assignment of variables to obtain solutions $(a = 1, b = c = d = e = f' = 0)$, $(a = 0, b = -1, c = 1, d = 0, e = 1, f = 0)$ and $(a = b = 0, c = -1/2, d = 1/2, e = 0, f = 1)$, which yield the complete set

$$\pi = \left\{ \alpha, \frac{lk}{m}, t\sqrt{\frac{g}{l}} \right\}.$$

21

Step 3. The equation in Buckingham's theorem takes the form

$$f\left(\alpha, \frac{lk}{m}, t\sqrt{\frac{g}{l}}\right) = 0.$$

Step 4. Solve the equation in Step 3 for the final product to obtain

$$t\sqrt{\frac{g}{l}} = h\left(\alpha, \frac{lk}{m}\right),$$

or

$$t = \sqrt{\frac{l}{g}}\, h\left(\alpha, \frac{lk}{m}\right).$$

12. Answers to Model Exam

Step 1. $S = (l, v, \delta, \alpha, \rho, \mu, a, \phi)$.

Step 2.

l	v	δ	α	ρ	μ	a	ϕ
L	LT^{-1}	L	—	ML^{-3}	$ML^{-1}T^{-1}$	LT^{-1}	MLT^{-2}

The resulting system of equations is

$$\begin{cases} e + f \phantom{{}+ g} + h = 0 \\ a + b + c - 3e - f + g + h = 0 \\ -b \phantom{{}+ c - 3e} - f - g - 2h = 0. \end{cases}$$

$$\begin{bmatrix} 0 & 0 & 0 & 0 & 1 & 1 & 0 & 1 \\ 1 & 1 & 1 & 0 & -3 & -1 & 1 & 1 \\ 0 & -1 & 0 & 0 & 0 & -1 & -1 & -2 \end{bmatrix}$$

$$\sim \begin{bmatrix} 0 & 0 & 0 & 0 & 1 & 1 & 0 & 1 \\ 1 & 1 & 1 & 0 & 0 & 2 & 1 & 4 \\ 0 & -1 & 0 & 0 & 0 & -1 & -1 & -2 \end{bmatrix}$$

$$\sim \begin{bmatrix} 0 & 0 & 0 & 0 & 1 & 1 & 0 & 1 \\ 1 & 0 & 1 & 0 & 0 & 1 & 0 & 2 \\ 0 & 1 & 0 & 0 & 0 & 1 & 1 & 2 \end{bmatrix}.$$

Let c, d, f, g, h be the arbitrary variables. The assignments

$$(c = 1, d = f = g = h = 0), \quad (d = 1, c = f = g = h = 0),$$

$$(f = 1, c = d = g = h = 0), \quad (g = 1, c = d = f = h = 0),$$

$$(h = 1, c = d = f = g = 0)$$

yield the complete set

$$\left\{\frac{\delta}{l}, \alpha, \frac{\mu}{l v \rho}, \frac{a}{v}, \frac{\phi}{\rho v^2 l^2}\right\}.$$

Step 3.

$$f\left(\frac{\delta}{l}, \alpha, \frac{\mu}{l v \rho}, \frac{a}{v}, \frac{\phi}{\rho v^2 l^2}\right) = 0.$$

Step 4.

$$\phi = \rho v^2 l^2 h\left(\frac{\delta}{l}, \alpha, \frac{\mu}{l v \rho}, \frac{a}{v}\right).$$

UMAP Module 530

Modules in Undergraduate Mathematics and its Applications

Perturbation Theory

Bertram Ross

Published in cooperation with the Society for Industrial and Applied Mathematics, the Mathematical Association of America, the National Council of Teachers of Mathematics, the American Mathematical Association of Two-Year Colleges, The Institute of Management Sciences, and the American Statistical Association.

COMAP

"During the last 35 years, there has been an explosive interest in nonlinear differential equations. The purpose of this module is to provide an historically oriented expository introduction to one particular analytical method of solving certain nonlinear differential equations."

INTERMODULAR DESCRIPTION SHEET:	UMAP Unit 530
TITLE:	Perturbation Theory
AUTHOR:	Bertram Ross Department of Mathematics University of New Haven West Haven, CT 06156
REVIEW STAGE / DATE:	III 7/21/81
CLASSIFICATION:	Applied Calculus and Differential Equations/Analysis of Nonlinear Differential Equations
TARGET AUDIENCE:	Students in advanced calculus, applied mathematics, mathematical physics, mechanical and electrical engineering, and space engineering.
ABSTRACT:	This unit gives a historical and self-contained introduction to the method of solving nonlinear differential equations by expressing their solutions as power series expansions in powers of a small parameter. The basic example is the perturbation problem $u'' + u = \varepsilon u^2$. Computers cannot always obtain accurate solutions of equations like these when the parameter ε is very small; but one may approximate the solutions with truncated series in powers of ε. Students who read this unit learn an approach to the solution of nonlinear differential equations that is of great importance in mathematical physics and space engineering.
PREREQUISITES:	A standard calculus course, including infinite series and some exposure to differential equations.

© Copyright 1981, 1988 by COMAP, Inc. All rights reserved.

COMAP, 60 Lowell Street, Arlington, MA 02174 (617) 641-2600

Perturbation Theory

Bertram Ross
Department of Mathematics
University of New Haven
West Haven, CT 06516

Table of Contents

1. Introduction 1
2. Asymptotic Series 1
3. Introductory Example and Comments on Asymptotics . . . 2
4. Nonlinear Differential Equations and the
   Perturbation Problem 5
5. Example of a Regular Perturbation Problem 7
6. The Problem of Extending a Region of Validity 9
7. Concluding Remarks 12
8. Test Problem and Commentary 13
9. References 16
10. Annotated Bibliography 16
11. Answers and Solutions to Exercises 18
    Appendix A 21
    Appendix B 22

Modules and Monographs in Undergraduate Mathematics and Its Applications (UMAP) Project

The goal of UMAP was to develop, through a community of users and developers, a system of instructional modules in undergraduate mathematics and its applications to be used to supplement existing courses and from which complete courses may eventually be built.

The Project was guided by a National Advisory Board of mathematicians, scientists, and educators. UMAP was funded by a grant from the National Science Foundation and is now supported by the Consortium for Mathematics and Its Applications (COMAP), Inc., a non-profit corporation engaged in research and development in mathematics education.

COMAP Staff

Paul J. Campbell	Editor
Solomon A. Garfunkel	Executive Director, COMAP
Laurie W. Aragon	Business Development Manager
Philip A. McGaw	Production Manager
Theresa Cronin	Copy Editor
Annemarie S. Morgan	Administrative Assistant
John Gately	Distribution

The Project would like to thank S.J. Donnellan of IBM, Erwin Eltze of Fort Hays State University, William Glessner of Spinner Data Service, and Raymond Greenwell of Albion College for their reviews, and all others who assisted in the production of this unit.

This material was prepared with the partial support of National Science Foundation Grant No. SED80-07731. Recommendations expressed are those of the author and do not necessarily reflect the views of the NSF or the copyright holder.

A portion of this work has been previously printed in the *International Journal of Mathematical Education in Science and Technology*, Leicestershire, England.

1. Introduction

In the middle of the 20th century, the development of computers brought the promise of unlocking secrets in nonlinear problems that had previously baffled even the most skillful analysts. For example, until recently astronomers had been struggling for more than 250 years with the nonlinear systems of equations for planetary motion, making substantial progress only by finding clever transformations to linearize the problems. Computers now enable astronomers to make progress where none was possible before.

The advent of the computer, with its powerful new capabilities, has resulted in recent emphasizing of numerical approaches to many problems. However, we must not disregard the analytical method in favor of the computer. The classification of nonlinear differential equations into appropriate categories and the classification of the singular points of the solutions can only be done through an analytical approach [Ince 1956, Davis 1960].

During the last 35 years, there has been an explosive interest in nonlinear differential equations. The purpose of this module is to provide an historically oriented expository introduction to one particular analytical method of solving certain nonlinear differential equations. Series expansions in powers of a small parameter are used in the process of reducing these nonlinear differential equations to sequences of ordinary differential equations. An understanding of the techniques involved requires a background of ordinary differential equations and some exposure to infinite series. The series expansions in question, called *asymptotic series*, turn out to be different from the infinite series usually studied in calculus.

2. Asymptotic Series

"Certain expansions have been found useful even though the series diverge."

Certain expansions of functions in an infinite series, called asymptotic by Poincaré and semiconvergent by Stieltjes, have been found useful even though the series diverge. In asymptotic series expansions, prescribed accuracy is achieved by fixing the number of terms and by taking sufficiently small values of a parameter that may either appear naturally in an equation or be artificially introduced. In contrast, prescribed accuracy is achieved for convergent series by taking a sufficiently large number of terms.

It is interesting that the aims of mathematicians and those applied scientific workers who use series for calculations are frequently poles apart. After Cauchy's time, convergence became one of the central concepts in analysis. As a result, many mathematicians considered the remainder terms of an infinite series far more important than a finite number of terms at the beginning. On the other hand, physicists, astronomers, and other applied scientific workers often use series whose first few terms decrease in magnitude rapidly, and they care little whether the series converge or diverge.

"Table-making provided one of the strongest stimuli...."

Table-making provided one of the strongest stimuli in the early development of asymptotic methods. It is curious that no mention of this fact is made, either in standard texts on asymptotic solutions of differential equations or in those on asymptotic expansions of functions defined as integrals.

The first known example of an asymptotic series was that associated with the gamma function, first developed by De Moivre and later completed by Stirling in 1730. The idea of what is now recognized as an asymptotic solution to a differential equation appeared in the works of Liouville in the first half of the 19th century. Poincaré, aware that certain divergent series were useful, was determined to discover the important features of these series. His interest is quite remarkable, because at the time it was the fashion, as a result of Cauchy's work on convergence, to banish divergent series.

In 1886 Poincaré formally formulated criteria for asymptotic series palatable to the taste of the modern mathematician. The formal mathematical definition can be found in some applications-oriented advanced calculus texts. As a result of this definition, mathematicians began to regard these series with more favor. All the work of mathematicians up to the second decade of the 20th century, however, revolved about asymptotic solutions to self-adjoint differential systems, of which the Sturm-Liouville and Legendre equations are examples, and to the asymptotic expansions of functions defined as integrals.

3. Introductory Example and Comments on Asymptotics

Expansion in powers of a small parameter has become a common analytic technique for obtaining approximate solutions to certain nonlinear differential equations. We exemplify the basic idea at an elementary level by first considering the algebraic equation

$$u = 2 - \varepsilon u^2, \tag{1}$$

where ε is a positive number. Solving the above by the quadratic formula, we get

$$u = \frac{-1 \pm \sqrt{1 + 8\varepsilon}}{2\varepsilon}. \tag{2}$$

For $\varepsilon = 0.1$ the solution for the positive square root is

$$u = 1.70820393\ldots.$$

Following Poincaré, we assume an approximate solution for u in a series of ascending powers of ε:

$$u = 2 + u_1\varepsilon + u_2\varepsilon^2 + u_3\varepsilon^3 + \cdots + u_n\varepsilon^n, \tag{3}$$

where the u_i's are to be determined. We shall explain in due course why the usual ellipsis dots after $u_n\varepsilon^n$ have been omitted. The series (3) is really an infinite series, but for the purposes of calculation the series must be truncated at some appropriate term. The lead term on the right is 2 so that (3) and (1) will be in agreement when $\varepsilon = 0$. We hope to obtain a good approximation to the solution for small values of ε by using an appropriate number of terms of (3).

The coefficients u_i are determined successively. The first step is to substitute (3) into (1):

$$2 + u_1\varepsilon + u_2\varepsilon^2 + u_3\varepsilon^3 + \cdots$$
$$= 2 - \varepsilon\left(2 + u_1\varepsilon + u_2\varepsilon^2 + u_3\varepsilon^3 + \cdots\right)^2. \tag{4}$$

After squaring the term on the right in parentheses, we will have

$$2 + u_1\varepsilon + u_2\varepsilon^2 + u_3\varepsilon^3 + \cdots$$
$$= 2 - 4\varepsilon - 4u_1\varepsilon^2 - \left(4u_2 + u_1^2\right)\varepsilon^3$$
$$- \left(4u_3 + 2u_1u_2\right)\varepsilon^4 - \cdots.$$

Although these expressions have infinitely many terms, they are equal only if the coefficients of like powers of ε are equal. When we equate the first nine coefficients, we find

$$u_1 \qquad\qquad\qquad = -4$$
$$u_2 = -4u_1 \qquad\qquad\qquad = +16$$
$$u_3 = -4u_2 - u_1^2 \qquad\qquad\qquad = -80$$
$$u_4 = -4u_3 - 2u_1u_2 \qquad\qquad\qquad = +448$$
$$u_5 = -4u_4 - 2u_1u_3 - u_2^2 \qquad\qquad\qquad = -2688$$

$$u_6 = -4u_5 - 2u_1u_4 - 2u_2u_3 \qquad = +16896$$
$$u_7 = -4u_6 - 2u_1u_5 - 2u_2u_4 - u_3^2 \qquad = -109824$$
$$u_8 = -4u_7 - 2u_1u_6 - 2u_2u_5 - 2u_3u_4 \qquad = +732160$$
$$u_9 = -4u_8 - 2u_1u_7 - 2u_2u_6 - 2u_3u_5 - u_4^2 = -4{,}978{,}688$$

Then the assumed approximate solution **(3)** is

$$u = 2 - 4\varepsilon + 16\varepsilon^2 - 80\varepsilon^3 + 448\varepsilon^4 - 2688\varepsilon^5 + 16896\varepsilon^6$$
$$- 109824\varepsilon^7 + 732160\varepsilon^8 - 4{,}978{,}688\varepsilon^9 + \cdots, \qquad (5)$$

where the ellipsis dots denote terms with higher powers of ε. When $\varepsilon = 0.1$, we get, using the ten terms of **(5)**

$$u = 2 - .4 + .16 - .08 + .0448 - .02688 + .016896$$
$$- .0109824 + .00732160 - .004978688$$
$$= 1.7065733\ldots. \qquad (6)$$

The error that results from this method can be found by subtracting the value obtained from the exact solution given by **(2)**; it is $0.00167\ldots$.

For many successive terms in this example, the powers of the small parameter will continue to counterbalance the growth of the coefficients so that each term will be less in magnitude than its predecessor. However, the terms ultimately begin to increase in magnitude, and hence the series will diverge. It is typical of asymptotic expansions that the terms decrease in magnitude up to some point and then begin to increase. To obtain our approximation, we fixed the number of terms; that is, the series was truncated arbitrarily after the tenth term. For the best approximation, an asymptotic series is truncated so that the first increasing term is not included. To emphasize this important feature, the ellipsis dots in **(3)** were omitted.

> "...*the terms decrease up to some point and then begin to increase.*"

As an example letting $\varepsilon = 0.16$ in **(5)** we get

$$u = 2 - .64 + .4096 - .32768 + .29360 - .2818572$$
$$+ .283467 - \cdots. \qquad (7)$$

If we exclude the last term, we get $u = 1.4536$; the exact result from **(2)** with $\varepsilon = 0.16$ is $u = 1.5935\ldots$.

It can be shown that the error in asymptotic expansions depends upon the number of terms used and the size of the argument. Using too many terms reduces the accuracy of the representation. Finding the appropriate term for the truncation of an asymptotic series is often a separate mathematical problem. Details can be found in [Arfken 1966, 221].

Exercise
1. Use the method of this section to solve the equation

$$u = 3 - \varepsilon u^2,$$

with $\varepsilon = 0.1$:
   a. First let

$$u = \text{const} + u_1\varepsilon + u_2\varepsilon^2 + u_3\varepsilon^3 + \cdots,$$

   and find the appropriate value for the constant;
   b. Determine the values of the first seven coefficients $u_1, u_2, \ldots, u_7$;
   c. Substitute the values found in b and 0.1 for ε in the expression for u;
   d. Check your result by the quadratic formula.

4. Nonlinear Differential Equations and the Perturbation Problem

During the first 25 years of the 20th century, papers appeared sporadically on asymptotic solutions of nonlinear differential equations. But it was the 1937 pioneering work of Krylov and Bogoliubov, who applied asymptotic methods to solve problems in nonlinear mechanics, that appears to have been the basis for a number of significant advances in the 1960s and 1970s. An essentially complete theory existed in 1937 for linear differential equations which possess solutions with singularities that are poles or branch points. Even in the cases where linear differential equations do have solutions that possess essential singularities, considerable progress has been made and a large body of information has been assembled. But for nonlinear differential equations, satisfactory information existed only for certain restricted types of equations, possibly because the theory of functions has been developed largely around classes of functions for which linearity is an essential feature.

Since the work of Krylov and Bogoliubov, who employed the ideas of Poincaré and the astronomer Anders Lindstedt, many significant improvements have been made in the last three decades of highly sophisticated technological developments. Many of the problems that confront the applied mathematician of today, for example, those involving systems governed by nonlinear differential equations, stem from this technology. A consequence of nonlinearity is that the powerful tool of superposition has been lost. The lack of an effective replacement makes the solution of nonlinear differential equations a formidable hurdle.

The applied scientific worker has three major problems:

(I) achieve a reasonable analytic formulation often called model of the physical system,

(II) derive meaningful approximations that serve as solutions to the equations that describe the process,

(III) make the result more useful by seeking ways to widen the region of validity of the solution.

Computers cannot always obtain accurate solutions of nonlinear differential equations when the parameters are very small. Perturbation methods have proven satisfactory; as a result, the modern applied mathematician's repertoire needs to include this technique. The word perturbation in the mathematical sense is derived from the word "perturbed," which is applied to a system that differs slightly from a known system. Perturbation problems have been classified into two categories: *regular* and *singular*. An approximate solution of a nonlinear ordinary differential equation of the form

$$\frac{d^2u}{dt^2} + u = \varepsilon u^2, \tag{8}$$

with the solution given as an asymptotic series expansion of ascending powers of the small parameter ε, gives rise to a regular perturbation problem. A more difficult nonlinear ordinary differential equation to solve by asymptotic methods is one of the form

$$\varepsilon \frac{d^2u}{dt^2} + u = u^2, \tag{9}$$

which gives rise to a singular perturbation problem.

The general perturbation problem is to find the solution when ε is small, given the solution when $\varepsilon = 0$. If the order of the equation and the number of boundary conditions remain fixed in the procedure of solving the problem, it is called a regular perturbation problem, as in (8). However, if the order of the equation is reduced when $\varepsilon = 0$, and if one or more boundary conditions have to be discarded, the problem is called a singular perturbation problem, as in (9). Singular perturbation problems arise in boundary layer theory and in analyzing the edge effect in elasticity and plasticity.

5. Example of a Regular Perturbation Problem

Here is a detailed solution of a regular perturbation problem [Struble 1962, 63–71]. In the investigation of the orbital motion of the planets under the assumptions of general relativity, that is, the problem of the perihelion shift, Albert Einstein was led to solve the equation

$$\frac{d^2u}{dt^2} + u = a + \varepsilon u^2$$

by means of elliptic functions. Although we are concerned here only with solving this equation, we should not overlook the fact that modeling, the formulation of an equation that is a reasonable description of a physical system, is a difficult hurdle for the applied mathematician to overcome.

We will consider the case $a = 0$. We therefore have

$$\frac{d^2u}{dt^2} + u = \varepsilon u^2, \tag{10}$$

where ε is a small positive number. The initial conditions are:

at $t = 0$, $u = A$ (a constant); and at $t = 0$, $du/dt = 0$. \qquad (11)

In functional notation this is written

$$u(0) = A,\ u'(0) = 0.$$

First we observe that when $\varepsilon = 0$, (10) has the solution

$$u = A \cos t. \tag{12}$$

Let the assumed approximate solution be of the form

$$u = u_0(t) + u_1(t)\varepsilon + u_2(t)\varepsilon^2 + \cdots, \tag{13}$$

where the coefficients u_i are to be determined and the ellipsis dots denote terms in higher powers of ε. The right side of (13) is substituted into (10) in the same manner that (3) was substituted into (1). Coefficients of like powers of ε are then equated. We get a sequence of linear ordinary differential equations, which are solved successively:

$$\frac{d^2u_0}{dt^2} + u_0 = 0,$$

$$\frac{d^2u_1}{dt^2} + u_1 = u_0^2,$$

$$\frac{d^2u_2}{dt^2} + u_2 = 2u_0u_1,$$

$$\vdots \tag{14}$$

The solution of the first of the equations above is substituted into the second. The solutions of the first and second equations are substituted ito the third, and so on. Such a sequence is sometimes called a *nested sequence*.

From **(12)** we see that $u_0 = A \cos t$. Then the second equation in **(14)** is

$$\frac{d^2u_1}{dt^2} + u_1 = A^2 \cos^2 t = \frac{A^2}{2} + \frac{A^2}{2} \cos 2t. \tag{15}$$

The first equation in **(14)** uses the initial conditions **(11)**. The remaining equations in **(14)** use the initial conditions $u_i(0) = 0$ and $u_i'(0) = 0$ (see Appendix A for the reason). Solving **(15)** with $u_1(0) = 0$ and $u_1'(0) = 0$ we get

$$u_1 = A^2\left(\tfrac{1}{2} - \tfrac{1}{3}\cos t - \tfrac{1}{6}\cos 2t\right). \tag{16}$$

The third equation in **(14)** is

$$\frac{d^2u_2}{dt^2} + u_2 = 2u_0u_1. \tag{17}$$

The values of u_0 and u_1 obtained in **(12)** and **(16)** are substituted in the right side above. When the zero initial conditions are applied, we get

$$u_2 = A^3\left(-\tfrac{1}{3} + \tfrac{29}{144}\cos t + \tfrac{5}{12}t\sin t + \tfrac{1}{9}\cos 2t + \tfrac{1}{48}\cos 3t\right). \tag{18}$$

Thus, the solution to our problem as given by **(13)** is

$$u = A\cos t + A^2\left(\tfrac{1}{2} - \tfrac{1}{3}\cos t - \tfrac{1}{6}\cos 2t\right)\varepsilon$$
$$+ A^3\left(-\tfrac{1}{3} + \tfrac{29}{144}\cos t + \tfrac{5}{12}t\sin t\right.$$
$$\left. + \tfrac{1}{9}\cos 2t + \tfrac{1}{48}\cos 3t\right)\varepsilon^2 + \cdots. \tag{19}$$

This solution does not suggest periodicity, because of the presence of the term $t \sin t$.

If ε and the product $A\varepsilon$ are sufficiently small in magnitude, the solution **(19)** can be shown to be an asymptotic approximation of u for a finite interval of t.

6. The Problem of Extending a Region of Validity

We are now ready to consider the third hurdle of the applied mathematician, the widening of the region of validity of the solution. Terms of the form

$$t^n \cos t \quad \text{and} \quad t^n \sin t, \tag{20}$$

one of which appeared as early as the third term in **(19)**, are called *secular* terms. The independent variable t multiplies the harmonic terms and they become unboundedly large as t grows large. Astronomers were disturbed by the intrusion of these non-periodic terms in the description of phenomena that were fundamentally periodic. They were called secular since they represented variations that progress in one direction for long periods of time, even though ultimately they may prove to be periodic. In astronomy the variations in t are relatively small, but in electrical systems the appearance of a term of the form $t^n \cos t$ would immediately introduce the phenomenon of resonance. It became, therefore, of great interest to make the result **(19)** more useful for computation for larger values of t.

It was probably Lindstedt in 1882 who first conceived the brilliant idea that a different form of the solution to **(10)** might be obtained by an appropriate change of variable. *Lindstedt's procedure*, as it is called, produces a delay in the appearance of the secular terms, which allows more terms in the series solution for computation for larger t and thus widens the region of validity of the solution. In the last few decades, other ingenious ideas have been devised to accomplish the same result. The following example shows the combination of Poincaré's and Lindstedt's contributions.

The assumed solution of **(10)** is written as

$$u = y_0(\omega t) + y_1(\omega t)\varepsilon + y_2(\omega t)\varepsilon^2 + \cdots, \tag{21}$$

where the y_i's, which are to be determined, will be required to be periodic functions of period 2π. Let ω be the true frequency and be determined by the expansion

$$\omega = 1 + \omega_1 \varepsilon + \omega_2 \varepsilon^2 + \cdots. \tag{22}$$

The first term on the right is unity, so that when $\varepsilon = 0$, the frequencies in (22) and (19) will be in agreement.

The expansions (21) and (22) taken together represent a reorganization of the terms in (19) in such a way as to produce periodic coefficients for the power series (21). The true frequency ω is introduced because the nonlinearity in (10) not only affects amplitude but unexpectedly affects the frequency.

Letting $\omega t = \tau$, we transform (10) into

$$\omega^2 \frac{d^2 u}{d\tau^2} + u = \varepsilon u^2. \tag{23}$$

(See Appendix B for details.) From (21) and (22) we get

$$\frac{d^2 u}{d\tau^2} = \frac{d^2 y_0}{d\tau^2} + \frac{d^2 y_1}{d\tau^2} \varepsilon + \frac{d^2 y_2}{d\tau^2} \varepsilon^2 + \cdots,$$

$$\varepsilon u^2 = y_0^2 \varepsilon + 2 y_0 y_1 \varepsilon^2 + \left(2 y_0 y_2 + y_1^2\right) \varepsilon^3 + \cdots,$$

$$\omega^2 = 1 + 2\omega_1 \varepsilon + \left(2\omega_2 + \omega_1^2\right) \varepsilon^2 + \cdots. \tag{24}$$

Eqs. (24) are substituted into (23) and regrouped into powers of ε. When coefficients of like powers of ε are equated, we obtain a sequence of linear differential equations that can be solved successively:

$$\frac{d^2 y_0}{d\tau^2} + y_0 = 0$$

$$\frac{d^2 y_1}{d\tau^2} + y_1 = y_0^2 - 2\omega_1 \frac{d^2 y_0}{d\tau^2}$$

$$\frac{d^2 y_2}{d\tau^2} + y_2 = 2 y_0 y_1 - \left(\omega_1^2 + 2\omega_2\right) \frac{d^2 y_0}{d\tau^2} - 2\omega_1 \frac{d^2 y_1}{d\tau^2}$$

$$\vdots \tag{25}$$

From the first of Eqs. (25), we see that $y_0 = A \cos \tau$, from which it follows that $d^2 y_0 / d\tau^2 = -A \cos \tau$. The second of Eqs. (25), with the use of a trigonometric identity, can then be written as

$$\frac{d^2 y_1}{d\tau^2} + y_1 = \frac{A^2}{2} + 2\omega_1 A \cos \tau + \frac{A^2}{2} \cos 2\tau. \tag{26}$$

The solution of **(26)** will have a secular (or resonance) term, because the term $\cos \tau$ duplicates a term in the homogeneous solution of the left side of **(26)**. To eliminate this possibility, the term with $\cos \tau$, in Lindstedt's words, is "cast out," by setting $\omega_1 = 0$. After applying zero initial conditions, the solution is

$$y_1 = A^2 \left(\tfrac{1}{2} - \tfrac{1}{3} \cos \tau - \tfrac{1}{6} \cos 2\tau \right), \tag{27}$$

which is periodic in τ of period 2π, as required.

The third of Eqs. **(25)**, with the use of a trigonometric identity, becomes

$$\frac{d^2 y_2}{d\tau^2} + y_2 = \left(\tfrac{5}{6} A^3 + 2A\omega_2 \right) \cos \tau$$

$$- A^3 \left(\tfrac{1}{3} + \tfrac{1}{3} \cos 2\tau + \tfrac{1}{6} \cos 3\tau \right). \tag{28}$$

The solution of **(28)** will have a secular term, unless ω_2 is chosen to make the quantity in parentheses in the first term on the right zero. Thus, $\omega_2 = -5A^2/12$. The solution of **(28)**, after applying zero initial conditions, is

$$y_2 = A^3 \left(-\tfrac{1}{3} + \tfrac{29}{144} \cos \tau + \tfrac{1}{9} \cos 2\tau + \tfrac{1}{48} \cos 3\tau \right). \tag{29}$$

In a similar manner, $y_3, y_4, \ldots$ can be obtained. Our solution from **(21)** has no secular terms and is

$$u = A \cos \tau + y_1 \varepsilon + y_2 \varepsilon^2 + \cdots, \tag{30}$$

where y_1 and y_2 are given by **(27)** and **(29)**. The exact frequency is given by **(22)**:

$$\omega = 1 + 0 - \frac{5A^2}{12} \varepsilon^2 + \cdots. \tag{31}$$

In the process of developing a series solution in which the appearance of secular terms can be delayed as far as desired accuracy requires, we have also found an expression for the true frequency.

Note on Singular Perturbations

To get a clearer understanding of what makes the singular perturbation problem more difficult to handle than the regular one, consider

$$\varepsilon \frac{d^2 u}{dx^2} + u = u^2, \tag{32}$$

where ε is a small positive number and where the boundary conditions are

$$\text{at } x = 0, u = 0 \quad \text{and at} \quad x = 1, u = 2. \tag{33}$$

In functional notation this is written: $u(0) = 0$ and $u(1) = 2$.

The order of the equation is reduced in the limiting case when $\varepsilon = 0$. The reduced equation has the solutions $u(x) = 0$ and $u(x) = 1$. Because the boundary conditions cannot both be satisfied, one boundary condition must be dropped.

Eq. (32) cannot have a straightforward solution of the form $u(x) = u_0(x) + u_1(x)\varepsilon + u_0(x)\varepsilon^2 + \cdots$, because when this assumed solution is substituted into (32), the leading equation, after equating coefficients of like powers of ε, will be $u_0 - u_0^2 = 0$. Here again, the boundary conditions cannot both be satisfied.

To overcome this difficulty, certain procedures called *inner expansions* and *outer expansions* have been developed. The region of validity of the solution is the region in an overlapping domain which both expansions are valid.

7. Concluding Remarks

"*...most problems are never 'completely solved.'*"

"*...many of the formal procedures are as yet mathematically unjustified.*"

You have undoubtedly observed by this time that the terms "solve" and "solution" have a degree of uncertainty about them. In the words of Poincaré, most problems are never "completely solved." In the case of differential equations, the reduction of the solution to a function contained in the classical corpus of functions is usually considered a highly satisfactory achievement. In other cases, the expression of a solution in the form of an infinite series suffices. In practical applications, however, there are those who will not be satisfied until the solution is reduced to a form where a computation can be obtained within a specified range of accuracy. Others are interested in the solution insofar as critical information can be obtained, such as periodicity and the locations of zeros and singularities. Because this paper is intended to be suggestive and heuristic, and specifically oriented for those with limited mathematical experience, we did not show, for example, that (30) is indeed a solution to (10). But the alternatives are either to accept a mathematically incomplete job or to deny a large audience the knowledge of this vital topic. It is important to note that many of the formal procedures in perturbation theory are as yet mathematically unjustified. Much research is going on in this area; the situation is reminiscent of Heaviside's work, which took nearly two decades to justify.

"Perturbation methods have certain advantages over numerical methods in the presence of a small parameter."

Perturbation methods, as mentioned earlier, have certain advantages over numerical methods in the presence of a small parameter. While numerical methods can deal with a fixed set of parameters, analytical methods show the dependence of the phenomena on the parameters. Even with a fixed set of parameters, analytical methods are useful for computation. For instance, they can predict the thickness of a boundary layer and assist in the refinement of mesh size (mesh is the subdivision of a rectangular plane domain of integration into rectangular grids).

The method of perturbations is applicable to the finding of approximate solutions of equations other than algebraic and certain nonlinear differential equations: e.g., partial differential equations, integral and functional differential equations. The earliest use of perturbation methods was made by the early astronomers in celestial mechanics, and in the first half of the twentieth century the emphasis was in the field of nonlinear mechanics. In the last decade, the fields of application broadened dramatically. An exciting feature of perturbation problems is its unexpected appearance in many diverse areas. Some of these fields are chemical reactions, diffusion, fluid mechanics, epidemiology and population studies. For more of these, see the references.

8. Test Problem and Commentary

Solve the differential equation

$$\frac{d^2y}{dt^2} + \omega^2 y = \varepsilon y^3, \tag{1}$$

which describes the oscillation of a mass on a string with a weakly nonlinear restoring force. The initial conditions are $y(0) = 1$ and $y'(0) = 0$.

Find an asymptotic solution in powers of ε:

$$y(t; \varepsilon) = y_0(t) + y_1(t)\varepsilon + \cdots, \tag{2}$$

and determine only y_0 and y_1.

Hints: Substitute (2) into (1) and equate coefficients of like powers of

ε and show that

$$\frac{d^2 y_0}{dt^2} + \omega^2 y_0 = 0, \tag{3}$$

and

$$\frac{d^2 y_1}{dt^2} + \omega^2 y_1 = y_0^3. \tag{4}$$

After solving for y_0 in **(3)** using the given initial conditions, and substituting for y_0 in **(4)**, show that

$$\frac{d^2 y_1}{dt^2} + \omega^2 y_1 = \cos^3 \omega t. \tag{5}$$

Use the identity

$$\cos^3 \omega t = (3/4)\cos \omega t + (1/4)\cos 3\omega t$$

to show that the general solution is the sum of the homogeneous (complementary) solution and the particular solution:

$$y_1 = c_1 \cos \omega t + c_2 \sin \omega t - \frac{1}{32\omega^2}\cos 3\omega t + \frac{3t}{8}\sin \omega t. \tag{6}$$

Impress zero initial conditions on **(6)** to solve for c_1 and c_2. The solution as given by **(2)** is

$$y = \cos \omega t + \left[\frac{1}{32\omega^2}(\cos \omega t - \cos 3\omega t) + \tfrac{3}{8}t \sin \omega t\right]\varepsilon + \cdots .$$

"... due to the appearance of secular terms, the solution cannot be carried out to sufficiently large t ..."

The term $t \sin \omega t$ is a secular term; therefore, the expansion above will not be valid for large t, even though it may be suitable for small t. The solution is periodic; however, due to the appearance of secular terms, the solution cannot be carried out to sufficiently large t to calculate the distortion of the period stemming from the nonlinear restoring force.

The appearance of secular terms in the solution may sometimes be delayed by appropriate changes of variable.

Exercise

2. The purpose of this exercise is to demonstrate the difficulties of ordinary computation in the presence of a small parameter. The

three steps of perturbative analysis and the power of an expansion in powers of a small parameter are exemplified.

a. Solve for the negative root of $u^3 - 4.001u + .002 = 0$ using Newton's method, $u_{n+1} = u_n - f(u_n)/f'(u_n)$.

b. Compare the work and result above when the same equation is solved by the introduction of a small parameter.

Step i: Convert the original problem into a perturbation problem.

$$u^3 - (4 + \varepsilon)u + 2\varepsilon = 0, \tag{1}$$

which reproduces the original with $\varepsilon = 0.001$.

Step ii: Assume a series solution for u in powers of a small parameter.

$$u = -2 + u_1\varepsilon + u_2\varepsilon^2 + \cdots . \tag{2}$$

The lead term is obtained by setting $\varepsilon = 0$ in (1). Substitute (2) into (1); and for the purpose of this example, neglect terms with powers of ε greater than 2. Regroup terms with like powers of ε, obtaining

$$(-8 + 8) + (8u_1 + 4)\varepsilon + (8u_2 - u_1 - 6u_1^2)\varepsilon^2 + \cdots = 0.$$

Set each coefficient of ε in parentheses equal to zero. The result is

$$u = -2 - \tfrac{1}{2}\varepsilon + \tfrac{1}{8}\varepsilon^2 + \cdots . \tag{3}$$

Step iii: Obtain the answer to the original problem by setting $\varepsilon = 0.001$ in (3).

Exercises

3. Given the nonlinear ordinary differential equation

$$\frac{dy}{dx} - y^2 = x^2$$

with the initial condition: at $x = 0$, $y = 1$.

a. Introduce the parameter ε as a coefficient of the nonlinear term and find the solution in a series expansion in powers of ε.

b. Solve by taking five implicit differentiations: $y'(x) - y^2 = x^2$, $y'' - 2yy' = 2x$, $y''' - 2yy'' - 2(y')^2 = 2$, etc. Show $y'(0) = 1$, $y''(0) = 2$, $y'''(0) = 8$, $y^{(iv)}(0) = 28$, $y^{(v)} = 144$.

With these values of $y^{(n)}(0)$ as coefficients of a Taylor series, solve for $y(x)$.

c. Let $\varepsilon = 1$ in the solution of part (a). Show then that the solutions in part (a) and part (b) are the same.

4. G. Duffing (1918), in an extensive investigation of forced vibrations, dealt with an equation of the form

$$\frac{d^2u}{dt^2} + u = -\varepsilon u^3. \tag{i}$$

a. Although (i) can be solved by elliptic functions, the problem here is to solve it by perturbation methods. The initial conditions are $u(0) = 1$ and $u'(0) = 0$; that is, at $t = 0$, $u = 1$ and $du/dt = 0$.

b. In the solution to part (a), a secular term appears in the second term. Delay this appearance one term. Also, determine the true frequency. Hint: Let $\omega t = \tau$.

9. References

Annotated Bibliography

Ames, William F. 1965. *Nonlinear Partial Differential Equations in Engineering*. New York: Academic.

A good explanation of the concept of nonlinearity is given on pp. 2–3. Perturbation concepts start on p. 197, and an example is given of a solution in powers of a small parameter that originates from vibration theory. Ames explains the requirements of zero initial conditions that are applied after the first of the nested linear differential equations. (See Appendix A of this module.) Singular perturbations start on p. 211.

Arfken, George. 1966. *Mathematical Methods for Physicists*, New York: Academic. 1962.

Bellman, Richard. 1964. *Perturbation Techniques in Mathematics, Physics and Engineering*. New York: Holt, Rinehart and Winston.

Excellent but too succinct for real beginners. Parts I and II deal with expansions in powers of a small parameter. Of special interest to some is the development of Lagrange expansions, useful in the applications of series reversion. Of interest to this particular module is Section 6, an outline of methods to delay the appearance of secular terms.

Bender, Carl M., and Steven A. Orszag. 1978. *Advanced Mathematical Methods for Scientists and Engineers*. New York: McGraw-Hill.

Excellent but does not emphasize special methods nor dwell on equations whose exact solutions are known. The mathematical methods discussed are known collectively as asymptotic and perturbative analysis. Regular and singular perturbation problems are given, with thorough, detailed explanations.

Cole, Julian D. 1968. *Perturbation Methods in Applied Mathematics*. Waltham, MA: Blaisdell.

Written from a point of view of an applied mathematician. Less attention is paid to rigor than to rooting out fundamental ideas. The first section deals with developing a background for asymptotic expansions. The second and third sections present numerous examples of techniques. The later sections present numerous physical examples. This text assumes the reader to have a strong background in classical analysis.

Davis, Harold T. 1960. *Introduction to Nonlinear Differential and Integral Equations*. Washington, D.C.: United States Atomic Energy Commission.

Erdélyi, Arthur. 1976. Singular perturbations. In *Trends in Applications of Pure and Applied Mathematics*, edited by G. Fishera, 53–62. San Francisco: Pitman.

Surveys various problems involved in perturbation techniques, such as the proper classification of problems as regular or singular, matching, existence, and the circumstances under which the limiting problem has a unique solution.

Goldstein, Marvin, E., and Willis H. Braun. 1973. *Advanced Methods for the Solution of Differential Equations*. Washington, D.C.: National Aeronautics and Space Agency, NASA-SP 316.

Certain modern ideas, such as the theory of matched asymptotic expansions, which have not found their way into most conventional texts, are included. The level of presentation is at the graduate level but this is outweighed by a pleasant expository manner.

Ince, E. L. 1956. *Ordinary Differential Equations*, New York: Dover.

Nayfeh, Ali Hassan. 1973. *Perturbation Method*. New York: Wiley.

Straightforward solutions in powers of a parameter have limited regions of validity. This text deals in part with the problems of rendering these expansions uniformly valid, pointing out the advantages and limitations of different techniques. The perturbation problems are drawn from physics and engineering. The Lindstedt-Poincaré technique of delaying secular terms is explained in detail, together with other methods. The approach in this text is heuristic rather than rigorous.

O'Malley, Robert E. 1974. *Introduction to Singular Perturbations*. New York: Academic.

Gives a short section on regular perturbations and devotes the rest of the text to singular perturbation theory and problems. The author is also the editor of the *Proceedings of the Symposium on Asymptotic Methods and Singular Perturbations*. (Providence, RI: American Mathematical Society, 1976), a collection of problems drawn from physical situations.

Pipes, Louis A. 1965. *Operational Methods in Nonlinear Mechanics*. New York: Dover.

Excellent for those primarily interested in learning techniques and applications for the electrical engineer. No attempt is made to present a general theory. If Pipes dwells too much on solving the sequence of linear equations by the Laplace transform method, this is a small price to pay.

Struble, Raimond, A. 1962. *Nonlinear Differential Equations*, New York: McGraw-Hill.

Shohat, J. 1944. On Van der Pol's equations. *Journal of Applied Physics* 15:568–574.

Van Dyke, Milton. 1964. *Perturbation Methods in Fluid Mechanics*, New York: Academic.

Beginners will be interested in the first three chapters, in which the general ideas of asymptoticity are discussed. The examples are drawn exclusively from fluid mechanics but will be useful to workers in other fields interested in singular perturbation problems.

Wasow, Wolfgang. 1965. *Asymptotic Expansions for Ordinary Differential Equations*. New York: Interscience. Chapter III contains a good introduction to asymptotic power series. Chapter VII discusses expansions in powers of a parameter.

11. Answers and Solutions to Exercises

1. a. 3
   b. u_1 $\qquad = -9$
   $u_2 = -6u_1 \qquad = 54$
   $u_3 = -6u_2 - u_1^2 \qquad = -405$
   $u_4 = -6u_3 - 2u_1 u_2 \qquad = 3{,}402$
   $u_5 = -6u_4 - 2u_1 u_3 - u_2^2 \qquad = -30{,}618$
   $u_6 = -6u_5 - 2u_1 u_4 - 2u_2 u_3 \qquad = 288{,}684$
   $u_7 = -6u_6 - 2u_1 u_5 - 2u_2 u_4 - u_3^2 = -2{,}814{,}669$
   c. 2.56
   d. 2.42

3. a.

$$\frac{dy}{dx} - \varepsilon y^2 = x^2, \qquad y(0) = 1. \tag{1}$$

Assume the solution

$$y(x) = y_1(x) + y_2(x)\varepsilon + y_3(x)\varepsilon^2 + y_4(x)\varepsilon^3 + \cdots. \tag{2}$$

Substitute (2) into (1) getting

$$y_1' + y_2'\varepsilon + y_3'\varepsilon^2 + y_4'\varepsilon^3 + \cdots$$
$$-\varepsilon\left[\left(y_1^2 + 2y_1 y_2 \varepsilon + (2y_1 y_3 + y_2^2)\varepsilon^2\right.\right.$$
$$\left.\left.+ (2y_1 y_4 + 2y_2 y_3)\varepsilon^3 + \cdots\right] = x^2. \tag{3}$$

Equate like coefficients of ε and generate a sequence of ordinary linear differential equations, which are solved successively. In the first of these, apply the given initial conditions; and in the remainder, impress zero initial conditions.

$$\frac{dy_1}{dx} = x^2, \quad \text{which yields } y_1 = 1 + \frac{x^3}{3}.$$

$$\frac{dy_2}{dx} = 2y_1^2 = 2\left(1 + \frac{x^3}{3}\right)^2, \quad \text{which yields } y_2 = x + \tfrac{1}{6}x^4 + \tfrac{1}{63}x^7.$$

$$\frac{dy_3}{dx} = 2y_1 y_2, \quad \text{which yields } y_3 = x^2 + \frac{x^5}{5} + \frac{x^8}{56} + \frac{2x^{11}}{(11)(189)}.$$

$$\frac{dy_4}{dx} = 2y_1 y_3 + y_2^2, \quad \text{which yields } y_4 = \left(x^3 + \cdots\right).$$

The solution given by (2) is

$$y(x) = \left(1 + \frac{x^3}{3}\right) + \left(x + \frac{x^4}{6} + \frac{x^7}{63}\right)\varepsilon$$

$$+ \left(x^2 + \frac{x^5}{5} + \frac{x^8}{56} + \frac{2x^{11}}{(11)(189)}\right)\varepsilon^2$$

$$+ \left(x^3 + \cdots\right)\varepsilon^2 + \cdots. \tag{4}$$

b.

$$y' = x^2 + y^2, \; y'' = 2yy' + 2x, \; y''' = 2 + 2yy'' + 2(y')^2$$
$$y^{(iv)} = 6y'y'' + 2yy''', \; y^{(v)} = 6(y'')^2 8y'y''' + 2yy^{(iv)}.$$

Letting $x = 0$, we will have

$$y^{(0)} = 1, \quad y'(0) = 1, \quad y''(0) = 2, \quad y'''(0) = 8, \quad y^{iv}(0) = 28.$$

Using the values of $y^{(n)}(0)$ above as coefficients of the Taylor series expansion, we obtain

$$y(x) = 1 + x + x^2 + \tfrac{4}{3}x^3 + \tfrac{7}{6}x^4 + \cdots. \tag{5}$$

c. Let $\varepsilon = 1$ in (4) and collect like terms:

$$y(x) = 1 + \frac{x^3}{3} + \left(x + \frac{x^4}{6} + \frac{x^7}{63}\right) + \left(x^2 + \frac{x^5}{4} + \frac{x^8}{56} + \cdots\right)$$

$$+ \left(x^3 + \text{terms with higher powers}\right) + \cdots$$

$$= 1 + x + x^2 + x\tfrac{4}{3}x^4 + \cdots.$$

4. a.

$$\frac{d^2u}{dt^2} + u + \varepsilon u^3 = 0. \tag{i}$$

19

Assume a solution in the form of

$$u(t) = u_0(t) + u_1(t)\varepsilon + u_2(t)\varepsilon^2 + \cdots .\tag{ii}$$

Substitute **(ii)** into **(i)**. Equate coefficients of like powers of ε and obtain, as in **Exercise 1**, a sequence of linear equations.

$$\frac{d^2 u_0}{dt^2} + u_0 = 0, \quad \text{and} \quad u_0 = \cos t.$$

$$\frac{d^2 u_1}{dt^2} + u_1 = -u_0^3 = -\cos^3 t.$$

$$\vdots$$

Solving these, we get

$$u(t) = \cos t - \left[\tfrac{1}{32}\cos t - \tfrac{1}{32}\cos 3t + \tfrac{3}{8}t \sin t\right]\varepsilon + \cdots .$$

b. The term $(3/8)t \sin t$ is a secular term; it is the object of this part of the problem to delay its appearance. We follow the procedure of **(21)**. Make the change of variable $\omega t = \tau$. Introduce the true frequency ω by the expansion

$$\omega = 1 + \omega_1 \varepsilon + \omega_2 \varepsilon^2 + \cdots, \text{ so that}$$
$$\omega^2 = 1 + 2\omega_1 \varepsilon + \left(2\omega_2 + \omega_1^2\right)\varepsilon^2 + \cdots .$$

Assume the solution in the form of

$$u(\tau) = u_0(\tau) + u_1(\tau)\varepsilon + u_2(\tau)\varepsilon^2 + \cdots .$$

When the change of variable $\omega t = \tau$ and the assumed solution are substituted into **(i)**, we get, after equating coefficients of like powers of ε, the following sequence of linear differential equations:

$$\frac{d^2 u_0}{d\tau^2} + u_0 = 0 \quad \text{and} \quad u_0 = \cos \tau.$$

$$\frac{d^2 u_1}{d\tau^2} + u_1 + 2\omega_1 \frac{d^2 u_0}{d\tau^2} + u_0^3 = 0 \quad \text{or}$$

$$\frac{d^2 u_1}{d\tau^2} + u_1 = 2\omega_1 \cos \tau - \cos^3 \tau$$

$$= 2\omega_1 \cos \tau - ((3/4)\cos \tau + (1/4)\cos 3\tau)$$

$$= (2\omega_1 - (3/4))\cos \tau - (1/4)\cos 3\tau.$$

The homogeneous solution for u_1 duplicates the $\cos \tau$ term on the right side of the above. To avoid this, let $\omega_1 = 3/8$. Then, subject to the conditions $u_1(0) = u_1'(0) = 0$, we will have

$$u_1 = \tfrac{1}{32}\cos 3t.$$

The secular term has now been delayed one term. We have

$$u(\tau) = \cos\tau - \left(\tfrac{1}{32}\cos 3\tau\right)\varepsilon + O(\varepsilon^2).$$

where $\tau = \omega t$.
The frequency is

$$\omega = 1 + \tfrac{3}{8}\varepsilon - \tfrac{15}{256}\varepsilon^2 + \cdots,$$

where $-15/256$ was obtained from the differential equation for u_2 by requiring the coefficient of the secular-causing term to be zero.

The term $O(\varepsilon^2)$ in the solution for $u(\tau)$ means that, for some fixed value of τ, the error between $u(\tau)$ and the sum of the first two terms on the right side of the equation for $u(\tau)$ is at most of the order of ε^2, as ε tends to zero through positive values.

Appendix A

The reason for impressing zero initial conditions on all the differential equations after the first is explained as follows:

Let the initial conditions of (10) be $u(0) = A$ and $u'(0) = B$, where A and B are constants that do not depend upon ε.

Let $u(t, \varepsilon) = u_0(t) + u_1(t)\varepsilon + u_2(t)\varepsilon^2 + \cdots$. We then have

$$A = u_0(0, \varepsilon) = u_0(0) + u_1(0)\varepsilon + u_2(0)\varepsilon^2 + \cdots, \tag{1}$$

and

$$B = u_0'(0, \varepsilon) = u_0'(0) + u_1'(0)\varepsilon + u_2'(0)\varepsilon^2 + \cdots. \tag{2}$$

Equating coefficients of like powers of ε on both sides of (1) and (2) gives

$$A = u_0(0),\ 0 = u_1(0),\ 0 = u_2(0), \ldots, \tag{3}$$

and

$$B = u_0'(0),\ 0 = u_1'(0),\ 0 = u_2'(0), \ldots. \tag{4}$$

(We let $B = 0$ in our example (10).)

Thus, except for the differential equation for u_0, all subsequent differential equations must satisfy zero initial conditions.

The solution ((19)), when written for $t = 0$, is

$$u = A + A^2\left(\tfrac{1}{2} - \tfrac{1}{3} - \tfrac{1}{6}\right)\varepsilon + A^3\left(-\tfrac{1}{3} + \tfrac{29}{144} + \tfrac{1}{9} + \tfrac{1}{48}\right)\varepsilon^2 + \cdots . \tag{5}$$

Because the sum of the terms in parentheses equals zero, all terms except the first vanish. If other than zero initial conditions were impressed on $u_1, u_2, \ldots$, there is little likelihood that all terms in **(5)** with ε would vanish at $t = 0$.

Appendix B

Scaling, the technique we use here, is the change of the argument of a function to reduce the complexity of solution of equations.

Given that $\omega t = \tau$, ω a constant, we show that $\omega^2 \, d^2u/d\tau^2 = d^2u/dt^2$.

$$\omega t = \tau. \tag{1}$$

Differentiation with respect to τ gives

$$\omega \, dt = d\tau. \tag{2}$$

Differentiation of (1) with respect to u yields

$$\omega \frac{dt}{du} = \frac{d\tau}{du}, \tag{3}$$

or

$$\frac{1}{\omega} \frac{du}{dt} = \frac{du}{d\tau}. \tag{4}$$

Now differentiate **(4)** with respect to τ:

$$\frac{1}{\omega} \frac{d}{d\tau}\left(\frac{du}{dt}\right) = \frac{d}{d\tau}\left(\frac{du}{d\tau}\right). \tag{5}$$

But $d\tau = \omega \, dt$ from **(2)**. Substitute this into **(5)**:

$$\left(\frac{1}{\omega}\right) \frac{d}{\omega \, dt}\left(\frac{du}{dt}\right) = \frac{d^2u}{d\tau^2}.$$

Finally,

$$\frac{d^2u}{dt^2} = \omega^2 \frac{d^2u}{d\tau^2}.$$

UMAP Module 576

Modules in Undergraduate Mathematics and its Applications

Randomized Response Technique: Getting in Touch With Touchy Questions

Paul Mullenix

Published in cooperation with the Society for Industrial and Applied Mathematics, the Mathematical Association of America, the National Council of Teachers of Mathematics, the American Mathematical Association of Two-Year Colleges, The Institute of Management Sciences, and the American Statistical Association.

COMAP

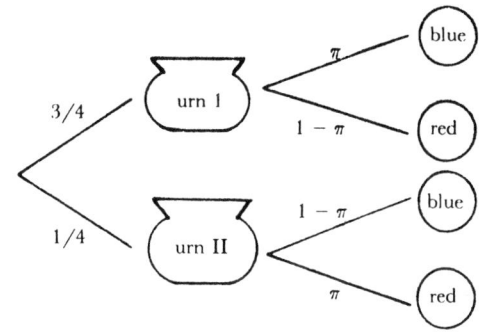

INTERMODULAR DESCRIPTION SHEET:	UMAP Unit 576
TITLE:	Randomized Response Technique: Getting in Touch with Touchy Questions
AUTHOR:	Paul Mullenix Harris Corporation Semiconductor Center, Statistical Development P.O. Box 883 Melbourne, FL 32901-0101
REVIEW STAGE / DATE:	III 7/30/81
CLASSIFICATION:	Statistics
TARGET AUDIENCE:	People in mathematics and social science courses that introduce conditional probability, or who are involved in questionnaire design.
ABSTRACT:	Intentionally falsifying answers or refusing to respond in a simple survey are common sources of bias. This is especially true when a topic in the survey is sensitive, e.g. estimating marijuana usage or abortion rates. This module explores a bias-reducing technique called "randomized response" in which a chance mechanism is used to preserve the anonymity of *particular* respondents while enabling the surveyor to discover information about the *group* of respondents by using elementary probability concepts. Students use the randomized response models of Warner, Simmons, and Greenberg to analyze data collected in surveys.
PREREQUISITES:	Conditional probability, confidence intervals (in one exercise only).

© Copyright 1981, 1988 by COMPAC, Inc. All rights reserved.

Randomized Response Technique: Getting in Touch With Touchy Questions

Paul Mullenix
Harris Corporation
Semiconductor Center,
Statistical Development
P.O. Box 883
Melbourne, FL 32901-0101

Table of Contents

1. INTRODUCTION 1
2. THE WARNER MODEL 3
3. THE SIMMONS MODEL 7
4. THE GREENBERG MODEL 12
5. CONCLUSION 15
6. FURTHER READINGS AND REFERENCES 18
7. ANSWERS TO EXERCISES 18

Modules and Monographs in Undergraduate Mathematics and Its Applications (UMAP) Project

The goal of UMAP was to develop, through a community of users and developers, a system of instructional modules in undergraduate mathematics and its applications to be used to supplement existing courses and from which complete courses may eventually be built.

The Project was guided by a National Advisory Board of mathematicians, scientists, and educators. UMAP was funded by a grant from the National Science Foundation and is now supported by the Consortium for Mathematics and Its Applications (COMAP), Inc., a non-profit corporation engaged in research and development in mathematics education.

COMAP Staff

Paul J. Campbell	Editor
Solomon A. Garfunkel	Executive Director, COMAP
Laurie W. Aragon	Business Development Manager
Philip A. McGaw	Production Manager
Theresa Cronin	Copy Editor
Annemarie S. Morgan	Administrative Assistant
John Gately	Distribution

This module was developed under the auspices of the UMAP Statistics Panel whose members were: Thomas R. Knapp, Chair, University of Rochester; Roger Carlson, University of Missouri; J. Richard Elliott, Wilfrid Laurier University; Earl Faulkner, Brigham Young University; David Herr, University of North Carolina; Peter Holmes, University of Sheffield; Peter Purdue, University of Kentucky; Judith Tanur, SUNY at Stony Brook; Maurice Tatsuoka, University of Illinois; and Douglas Zahn, Florida State University. The Project would like to thank the members of the Statistics Panel for their reviews, and all others who assested in the production of this unit.

This material was prepared with the partial support of National Science Foundation Grant No. SED80-07731. Recommendations expressed are those of the author and do not necessarily reflect the views of the NSF or the copyright holder.

1. Introduction

"Have you smoked marijuana in the past month?"

"Have you smoked marijuana in the past month?" "Do you belong to the Communist party?" "Did you cheat on your income tax return last year?" "Have you ever been unfaithful to your spouse?" These are examples of very sensitive questions. If such questions were asked directly in a sample survey, we would find that many people either refused to answer or lied in their answers. To social scientists, government officials or others, such questions provide useful information. But when people fear that their honesty may be used against them, how can the survey researcher discover the answers to sensitive questions and still respect the privacy of the respondent?

Many methods for gaining information about sensitive topics have been employed with varying degrees of success. Some psychological tests contain so-called "lie-items" to check for consistency of response, there have been special studies conducted to determine the extent of lying one may expect in a certain field, and unobtrusive measures have been used to obtain information indirectly as when increases in drug usage are compared with drug-related arrests or overdoses at a local hospital.

"This module presents an ingenious sample-survey method...."

This module presents an ingenious sample-survey method first used in 1965 by Stanley Warner. This method not only has the potential of providing reliable overall results, but it simultaneously respects the right to privacy of the interviewee. Warner termed the technique "randomized response," since a chance mechanism such as a die or a deck of cards determines which of a number of questions the respondent will answer, without revealing this choice to the interviewer. Others have since modified, improved, and extended randomized response models. After examining Warner's prototype in Section 2, in Sections 3 and 4 we will consider modifications proposed by W. R. Simmons and B. G. Greenberg.

Before introducing the Warner model, it may be helpful to review briefly some basic ideas from probability theory. The following section presents an elementary probability problem that strips away the background and setting of the Warner model. It illustrates the essentials of the calculations that will be necessary for solving problems with the randomized response technique.

1.1 An Urn Experiment

Consider the experiment depicted in **Figure 1**. First, an urn is selected—urn I is chosen with probability 3/4 and urn II is chosen with probability 1/4. Each urn contains blue and red balls (and no other colors), but we are not told how many balls are in each urn. We are told, however, that the proportion of blue balls in urn I, call it π, is the same as the proportion of red balls in urn II. Thus, because each urn contains only blue balls and red balls, we know that the proportion, $1 - \pi$, of red balls in urn I is equal to the proportion of blue balls in urn II (check **Figure 1**).

Second, we draw a ball from the urn initially selected.

If we repeat this experiment many times, finding that 7/20th of the time we end up drawing a blue ball, can we estimate the proportion, π, of blue balls in urn I?

To solve this problem, first note that we can draw a blue ball from either urn I or urn II. Thus,

$P(\text{draw a blue ball}) = P(\text{draw a blue ball from urn I})$
$\qquad\qquad\qquad\qquad + P(\text{draw a blue ball from urn II})$.

Now let B, A_1 and A_2 denote the events of drawing a blue ball, choosing urn I and choosing urn II respectively. Then the above probability statement may be written

$$P(B) = P(B \cap A_1) + P(B \cap A_2).$$

But, by the definition of conditional probability,

$$P(B \cap A_i) = P(A_i)P(B|A_i) \quad \text{for } i = 1, 2.$$

Hence, we have

$$P(B) = P(A_1)P(B|A_1) + P(A_2)P(B|A_2).$$

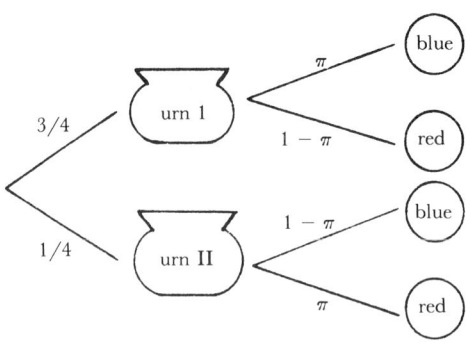

Figure 1. Tree diagram for the urn experiment.

The reader may recognize this as an instance of the law of total probability (see Exercise 2).

From **Figure 1** we see that $P(B|A_1) = \pi$, $P(B|A_2) = 1 - \pi$, $P(A_1) = 3/4$ and $P(A_2) = 1/4$. Since it was given that 7/20 of the time we choose a blue ball, $P(B) = 7/20$. Substituting these values yields

$$7/20 = 3/4\pi + 1/4(1 - \pi),$$
$$7/20 = 1/2\pi + 1/4,$$
$$1/10 = 1/2\pi,$$
$$1/5 = \pi.$$

Thus our estimate for π is 1/5, i.e., we estimate 1/5 of the balls in urn I are blue. (Since π is an *estimated value* in the above calculations, we will later denote this by the symbol $\hat{\pi}$.)

Exercises

1. If the experiment in this section is changed so that P(choosing urn I) = 4/5, P(choosing urn II) = 1/5, and we end up drawing a blue ball 1/2 the time, estimate the proportion of blue balls in urn I.

2. The law of total probability states: If $A_1, A_2, \ldots, A_n$ is a set of n mutually exclusive events with positive probabilities such that their union is the whole sample space, and B is any other event, then $P(B) = P(A_1)P(B|A_1) + P(A_2)P(B|A_2) + \cdots + P(A_n)P(B|A_n)$. Apply this to the following problem: An arcade contains three types of games: pinball, space-war games, and racing-car games. The games are designed so that an average player wins an extra play 2/5 of the time with pinball, 1/10 of the time with space-war games, and 1/4 of the time with racing-car games. If a person picks one of the three types of games at random, find the probability of winning an extra play.

2. The Warner Model

"The randomized response technique seeks to increase a subject's cooperation."

The randomized response technique suggested by Stanley Warner [1965] seeks to increase a subject's cooperation in a sample survey involving a sensitive topic. In the technique's simplest form, a randomizing device tells the subject which of two "yes" or "no" questions to answer. One of the questions deals with the sensitive issues, and it is the interviewer's desire to estimate what proportion answer "yes" to this sensitive question. The interviewer knows only the probability that the subject answers a particular question—and the probability of answering the sensitive question is not so high as to make the respondent skeptical or uneasy. Thus, for any *single* respondent, the interviewer cannot tell which question was answered. However, for a *group* of subjects, the interviewer may perform a probability calculation, exactly like that used in the urn experiment to estimate the proportion who respond "yes" to the sensitive question. By allowing the subject to respond to questions on a probability basis, individual privacy is maintained while information is revealed about the group.

> "...for any single respondent, the interviewer cannot tell which question was answered."

To illustrate the Warner model, suppose that we ask a group of 1000 politicians to cooperate in a study designed to estimate the prevalance of bribe acceptance. In order to persuade a politician to participate, we must guarantee that it will be impossible for anyone to determine whether or not a particular individual has accepted a bribe.

To do this, we instruct each politician in turn to sit at a table on which an ordinary deck of thoroughly shuffled playing cards has been spread faced down. To satisfy any doubts concerning the cards, the politician is allowed to examine the deck and reshuffle. The politician then selects one of the cards, privately examines the card, and returns it to the deck. Afterwards, the deck is again reshuffled to prevent the selected card from being identified. The politician is now directed to answer one of two questions, according to the suit of the card drawn. If the card is a heart, club, or a diamond, the yes-no question is (1): "At some time while holding public office, I have accepted money or favors which violate the legal or ethical codes pertaining to my office."

On the other hand, if the suit of the card is a spade, the yes-no question is (2): "At no time while holding public office, have I accepted money or favors which violate the legal or ethical codes pertaining to my office."

In this way, although the interviewer knows the politician's response, the interviewer does not know which question the politician answered. Since answers under this scheme cannot be incriminating, and the politician knows it, it is hoped that the reason for falsifying responses has been removed.

To see how the percentage of politicians who have accepted bribes may be estimated, suppose that this procedure is performed by each of the 1000 politicians, resulting in 300 "yes" answers. We now determine what proportion said "yes" to statement (1). According to the design of the experiment, a politician will respond to statement (1) whenever the card drawn is a heart, club or diamond. Let P be the probability that a politician responds to statement (1). Of the 52 cards in the deck, 39 are hearts, clubs or diamonds; therefore, the probability that the politician responds to statement (1) is $P = 39/52 = 3/4$. (The reader is asked to show in **Exercise 4** that the model has no solution if $P = 1/2$.) Similarly, the politician will reply to statement (2) with probability $1 - P = 1 - 3/4 = 1/4$. We could equivalently compute this probability directly: Since one will respond to statement (2) whenever a spade is drawn, the required probability is $13/52 = 1/4$.

If the politician answers "yes" to statement (1), then he or she is counted in the proportion (denoted π) who have taken bribes. But if the politician replied "yes" to statement (2), then he or she is counted in the $1 - \pi$ portion who have not accepted bribes. **Figure 2** summarizes the possible outcomes, where the symbols beside each path indicate probabilities.

The mathematical formulation of the model may now be obtained easily by comparing this problem with the urn experiment in Section 1.1. Urn I corresponds to statement (1) and urn II corresponds to statement (2). Likewise, drawing a blue ball in the urn experiment identifies with replying "yes" in this example.

The reasoning proceeds as in the urn experiment. In order for the subject to reply "yes," the subject must respond "yes" either to statement (1) or to statement (2). Hence we have

$$P(\text{"yes"}) = P(\text{"yes" on statement } (1))$$
$$+ P(\text{"yes" on statement } (2))$$
$$= P(\text{statement } (1) \text{ is chosen})$$
$$\times P(\text{"yes"}|\text{statement } (1) \text{ is chosen})$$
$$+ P(\text{statement } (2) \text{ is chosen})$$
$$\times P(\text{"yes"}|\text{statement } (2) \text{ is chosen}).$$

But $P(\text{"yes"}|\text{statement (1) is chosen})$ is the sought-after proportion π; this and the remaining quantities $1 - \pi$, P and $1 - P$ are

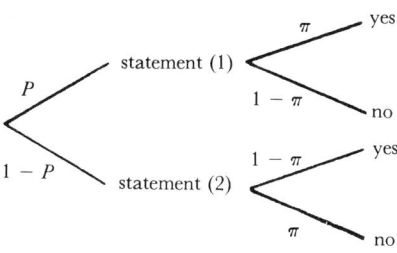

Figure 2. Tree diagram of the Warner model.

displayed in **Figure 2**. Thus,

$$P(\text{"yes"}) = P\pi + (1 - P)(1 - \pi),$$
$$= (39/52)\pi + (13/52)(1 - \pi),$$
$$= (3/4)\pi + (1/4)(1 - \pi),$$
$$= (1/2)\pi + 1/4.$$

Since 300 "yes" replies were recorded out of the 1000 trials, it follows that $P(\text{"yes"}) = 300/1000 = 0.30$. Substituting this into the above formula, and using the symbol $\hat{\pi}$ to denote the value used to estimate the population proportion π, gives

$$0.30 = 0.5\hat{\pi} + 0.25,$$
$$0.05 = 0.5\hat{\pi},$$
$$0.10 = \hat{\pi}.$$

Therefore we see that the percentage of those accepting bribes is estimated to be 10%. Quite remarkably, this estimation was achieved in such a way that it was impossible to identify any single individual who has accepted a bribe!

We now summarize the Warner approach as follows:

Warner Model

S1: I am a member of group A

S2: I am not a member of group A

$$P(\text{"yes"}) = P(S1)P(\text{"yes"}|S1) + P(S2)P(\text{"yes"}|S2)$$
$$P(\text{"yes"}) = P\pi + (1 - P)(1 - \pi), \text{ where}$$

$P =$ probability statement S1 is selected ($P \neq 1/2$)

$\pi =$ proportion of population who belong to group A.

Exercises

3. Solve the equation $P(\text{"yes"}) = P\pi + (1 - P)(1 - \pi)$ for π where $P = 5/6$ and $P(\text{"yes"}) = 5/8$.

4. Show that the Warner model has no unique solution in the case $P = 1/2$.

5. A research worker is assigned to estimate the size of the gay community in a large city. A total of 800 people are selected at random to be in the following experiment. The subject roles a pair of dice that are not visible to the interviewer. If the sum on

the dice is less than or equal to 9, the subject answers the question: "Do you belong to the gay community?" If the sum on the dice is greater than 9, the subject answers "yes" or "no" to, "Do you belong to the so-called 'straight' community?" If 190 "yes" answers are given, estimate the proportion of the population of the city accounted for by the gay community.

6. A march is planned in Southtown, U.S.A., with which the Ku Klux Klan is *not* sympathetic. Since the KKK is known to operate in the vicinity, the police chief wants to know whether to call in reserves to help keep order. The chief therefore wants to know the size of the KKK in Southtown and the surrounding area and so orders a quick study. A group of 1000 white males is selected at random. Each draws a card from a well-shuffled deck, privately examines the card, then returns the card to the deck. The deck is then reshuffled to conceal his selection. The interviewee is then instructed to answer:

"Do you belong to or actively support the KKK?" if the card drawn is a spade, club or diamond;
"Is it true that you do *not* belong to or actively support the KKK?" if the card drawn was a heart.

If 375 "yes" observations were made, estimate the density of KKK membership in the population. What is the population under discussion?

3. Simmons Model

The observant reader may have noticed that in the Warner model, either a "yes" or "no" answer may be construed as inculpatory. Suppose that belonging to a certain group A is perceived as being incriminating or perhaps embarrassing. Then using the above summary of the Warner model, where $S1$ and $S2$ are the two statements posed, a "yes" answer by the respondent would mean belonging to group A if the respondent had selected statement $S1$. Likewise a "no" answer would mean belonging to group A if statement $S2$ had been selected. Thus, either a "yes" or "no" answer may indicate that the respondent belongs to group A, the group with the sensitive characteristic.

To alleviate this, and thereby provide the subject with a greater sense of security, W. R. Simmons introduced a modification known as the "unrelated question" [Horvitz et al. 1967]. Just as in the Warner model, a subject chooses between two questions; however,

"...with the Simmons model only one of the questions is sensitive, the other is innocuous."

with the Simmons model only one of the questions is sensitive, the other is innocuous. It is hoped that the respondent will feel more comfortable with the unrelated question and have further reason to cooperate and respond truthfully.

For instance, in 1968 a study employing the Simmons model was reported by J. R. Abernathy et al. [1968]. The study attempted to estimate induced abortion rates among urban, North Carolina women between the ages of 18 and 44.

A *random sample* of about 1300 women in the childbearing ages of 18 to 44 was selected from five urban areas in North Carolina. Of these women, 1251 cooperated in the study; the others refused to be in the study, could not be located, were too ill to participate, or were otherwise unable to be included. (It should be noted that in order to use the randomized response technique, the group need *not* be a random sample. However, if one wishes to generalize from a sample to a population, then a random sample *is* necessary.)

First the interviewer chatted with the subject to establish some measure of trust and cooperation. The definition of an abortion as "an operation of some kind which a pregnant woman has in order to end her pregnancy and keep from having a baby, or something which she might do to herself to end the pregnancy and keep from having a baby" was also discussed. The entire survey procedure was thoroughly explained so that the subject was aware that her privacy would not be violated.

Each woman was presented with a small, sealed, transparent, plastic box containing 35 red balls and 15 blue balls. After shaking the box the subject tipped the box to one side, thereby allowing one of the balls to lodge in a window visible only to the respondent. If a red ball appeared in the viewer, she was instructed to respond to question (1): "Were you pregnant at some time during the past 12 months and had an abortion which ended the pregnancy?" On the other hand, if a blue ball appeared, she was asked to respond to question (2): "Were you born in the month of April?"

Note that question (2) is totally unrelated to the sensitive question involving abortion. It seems reasonable that with an unrelated question the subject might feel more secure that her response cannot be recognized and thus be even more willing to provide honest, straightforward replies. Continuing with the above procedure, 61 of the 1251 women responded "yes."

Now let P be the probability that the sensitive question is chosen, so that $1 - P$ is the probability that the unrelated question is selected. (Of course, P must not be so high as to arouse the suspicion of the respondent, i.e., she must be guaranteed that there is a reasonable probability that her response may be to the unrelated question.) In this study, $P = 35/50$ since there are 35 red balls and

15 blue ones. Define π_s to represent the proportion of women 18 to 44 in urban North Carolina who have had an abortion in the last year and set π_u equal to the proportion of women in the population born in April.

The probability of a "yes" response may now be derived from studying **Figure 3**.
Thus,

$$\begin{aligned} P(\text{"yes"}) &= P(\text{"yes" on question (1)}) \\ &\quad + P(\text{"yes" on question (2)}) \\ &= P(\text{question (1) is selected}) \\ &\quad \times P(\text{"yes"}|\text{question (1)}) \\ &\quad + P(\text{question (2) is selected}) \\ &\quad \times P(\text{"yes"}|\text{question (2)}) \\ &= P\pi_s + (1-P)\pi_u. \end{aligned}$$

Substituting the proportion of "yes" answers gives

$$61/1251 = (35/50)\hat{\pi}_s + (15/50)\hat{\pi}_u.$$

But before $\hat{\pi}_s$ may be found, one must know the value of $\hat{\pi}_u$, the estimate of the proportion of women in the population born in April.

To estimate the proportion π_u born in April, the birth records were checked of women for those years (1924–1950) when the 18- to 44-year-old women were born. In this way the figure 0.0826 was arrived at for $\hat{\pi}_u$. Substituting 0.0826 for $\hat{\pi}_u$ into the equation above and solving for $\hat{\pi}_s$ yields $\hat{\pi}_s = 0.0342$ for the estimate of the proportion who have had an abortion in the past year. Consequently, the estimate of the proportion of women in urban North Carolina having an abortion within the past 12 months was found to be 3%.

Notice that since the subjects in this experiment were randomly selected from the population, the number of "yes" answers follows a binomial distribution with parameter $\lambda = P(\text{"yes"})$. Recall that we

> "...the estimate of the proportion who have had an abortion in the past year...."

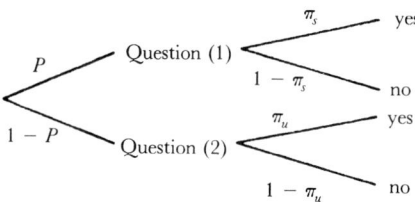

Figure 3. Tree diagram of the Simmons model.

estimate λ by computing the proportion of those in the *sample* who respond "yes" to the sensitive question. Using standard techniques (see Exercise 20 and its solution), we may derive a range of values in which the *population* value may be expected to lie. This range of values, called a *confidence interval* for λ, may then be used to find a confidence interval for the population value π_s. Confidence intervals tell us with a specified "confidence" that the true population value lies in this interval. For example, a 98% confidence interval for π_s in this abortion study is $(0.012, 0.057)$. This means that if we were to repeat this experiment many times (of course selecting a new sample each time), then in the long run about 98% of the confidence intervals for these experiments would contain the true population value of π_s. (For a fuller discussion of confidence intervals see [Bhattacharyya and Johnson 1977] or any standard statistics text.)

To see what the respondents in the abortion study thought of the randomized response technique, follow-up questions revealed that 67% of the subjects felt that their friends would not answer truthfully if the sensitive question were asked directly, 17% thought they would respond truthfully, and 16% were undecided. Furthermore, 60% of the women were confident that the other people would believe there was no trick to the box of balls which would allow the interviewer to identify the choice of question, 20% thought the contrary, and 20% were undecided.

We now summarize the Simmons model.

Simmons Model

$Q1$: sensitive question answerable by "yes" or "no"

$Q2$: unrelated question answerable by "yes" or "no"

$$P(\text{"yes"}) = P(Q1)P(\text{"yes"}|Q1) + P(Q2)P(\text{"yes"}|Q2)$$

$P(\text{"yes"}) = P\pi_s + (1 - P)\pi_u$, where

P = probability question $Q1$ is selected;

π_s = proportion of population answering "yes" to question $Q1$;

π_u = proportion of population answering "yes" to question $Q2$.

It may happen in the Simmons model that due to the choice of the unrelated question, π_u, the proportion who answer "yes" to the unrelated question, is unknown and no tables or records are available to easily estimate it. The parameter π_u may still be estimated, however—we may vary P, the probability of choosing the sensitive question, and repeat the experiment on a second group of subjects. One would then obtain the following pair of equations,

$$P(\text{"yes" in sample 1}) = P_1\pi_s + (1 - P_1)\pi_u$$
$$P(\text{"yes" in sample 2}) = P_2\pi_s + (1 - P_2)\pi_u.$$

This pair may then be solved simultaneously for π_s and π_u (see Exercise 10). Although certainly workable, this variation of the method has obvious drawbacks in the time, money and efficiency expended on the second sample.

Exercises

7. Solve the equation $P(\text{"yes"}) = P\pi_s + (1 - P)\pi_u$ for π_s given that $P = 7/10$, $\pi_u = 1/4$, and $P(\text{"yes"}) = 23/40$.

8. To determine the use of illegal stimulants by athletes at the Olympic games, suppose the International Olympic Committee selects 300 athletes at random and questions each individually. Each athlete picks a 2-digit number from a random number table, then is told to answer "yes" or "no" to:

    "I have knowingly used or plan to use illegal stimulants
      to enhance my performance," if the number is 00–79;
    "The digit I selected is odd," if the number is 80–99.

    If 65 "yes" responses were found, estimate the proportion that use illegal stimulants.

9. A study undertaken to determine the amount of alcohol consumed by high-school students was reported by Barth and Sandler [1976]. In the study 59 students were given 2 dimes (which they were allowed to keep) as a randomizing device. Upon the flipping of both coins, each was instructed to answer:

    "Does your telephone number end in an odd digit?" if
      both coins come up heads;
    "Over the past year have you consumed 50 or more
      glasses (or drinks) of any alcoholic beverages?" if the
      coins come up in any other combination.

    Given that 45 "yes" answers were recorded, what is the estimate of those who drank more than 50 glasses during the last year? Comment on the wording of the drinking question. Would someone *be able* to answer it accurately (consider the problem of memory and the meaning of a "glass").

10. Derive a formula for the solution of the Simmons model for unknown π_u. Let λ_1 and λ_2 be the probability of a "yes" response in the two samples, so that the model becomes

$$\lambda_1 = P_1 \pi_s + (1 - P_1) \pi_u,$$
$$\lambda_2 = P_2 \pi_s + (1 - P_2) \pi_u.$$

Show that the model has no solution for $P_1 = P_2$.

4. The Greenberg Model

Even though the unrelated question technique suggested by Simmons has proved to be a definite improvement over Warner's model, it is not the final state of the art. For if π_u is unknown, then two samples are necessary, a procedure that is costly at best. In 1969, however, B. G. Greenberg proposed that the response to the unrelated question be intrinsic to the randomizing device. Under Greenberg's model the randomizing device tells the subject to do one of three things:

(1) to answer the sensitive statement;
(2) to respond "yes";
(3) to respond "no".

Notice that in the second and third case, the subject is simply commanded to say a certain word—"yes" or "no"—the subject does not respond to a question at all. This procedure not only eliminates the need for two samples, but it also makes the computations very simple.

To illustrate this, suppose that due to recent publicity regarding marijuana usage among police recruits, an independent agency initiates a study to determine the proportion of recruits in a large police department who have recently smoked marijuana. Not fully trusting the reliability of the polygraph, the agency adopts the following plan.

Each recruit is presented with a black bag containing 100 numbered balls; other than numbering, the balls are identical. The recruit is allowed to inspect this randomizing device, then shake the bag and draw a ball. After personally examining the ball, the recruit replaces it in the bag and again shakes the bag to prevent the identification of the ball drawn. The recruit next responds "yes" or "no" according to the following scheme. If the ball is numbered 1–70, the recruit responds to the statement: "I have smoked or used marijuana in the past 30 days." If the ball is numbered 71–85, the recruit must answer "yes"; while if the ball is numbered 86–100, the recruit is instructed to answer "no."

As a consequence of this procedure, no one except the recruit knows whether or not the "yes" or "no" refers to the sensitive statement or, if by virtue of the number of his drawn ball, the recruit was forced to simply reply "yes" or "no" without regard to the sensitive statement. Moreover, the probability of the unrelated alternative is now simple to compute; a second sample is unnecessary, as is resorting to a table or other resources, to estimate the probability of the unrelated choice.

Suppose that of the 250 recruits subjected to the experiment, 62 answer "yes." The probability of a "yes" response under this Greenberg model is now easily derived (see **Figure 4**). We have

$$P(\text{"yes"}) = P(\text{"yes" and ball was numbered 1-70})$$
$$+ P(\text{"yes" and ball was numbered 71-85})$$
$$= P(\text{ball numbered 1-70})$$
$$\times P(\text{"yes"}|\text{ball numbered 1-70})$$
$$+ P(\text{ball numbered 71-85})$$
$$\times P(\text{"yes"}|\text{ball numbered 71-85})$$

$$62/250 = (70/100)\hat{\pi} + (15/100)(1)$$
$$0.098 = 0.7\hat{\pi}$$
$$0.14 = \hat{\pi}.$$

Thus the experiment estimates that about 14% of the recruits for this police department have recently used marijuana. (Since the third alternative of drawing a ball number 86–100 does not enter into the above calculations, the reader should consider why it is included in the model.)

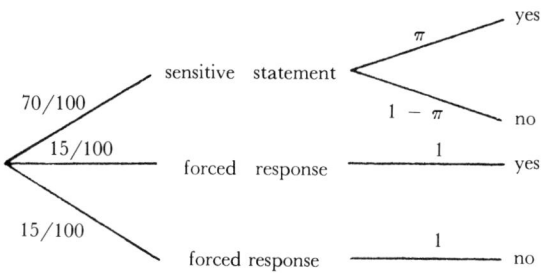

Figure 4. Greenberg model tree diagram for police recruitment example.

The Greenberg model may be summarized as follows:

Greenberg Model

A: sensitive statement

B: "yes"

C: "no"

$$P(\text{"yes"}) = P(A)P(\text{"yes"}|A) + P(B)P(\text{"yes"}|B)$$

$$P(\text{"yes"}) = P(A)\pi + P(B)$$

The probabilities of choosing A, B, and C are built into the randomizing device.

Exercises

11. Solve the equation $\lambda = p\pi + q$ for π if $p = 4/5$, $q = 1/10$ and $\lambda = 2/5$.

12. A federal investigator visits a religious commune deep in a tropical jungle, where it is suspected that some of the people are being held against their will. When he arrives he discovers that he is not allowed to question the inhabitants in private. Therefore, those who truly wish to leave cannot say so because of peer pressure. Fortunately, the investigator has had a course in probability and devises the following plan. He puts 70 dimes, 17 nickels and 15 quarters into a collection plate and mixes the coins. He is told that if he donates the $11.50 he may continue. He then covers the plate with a black cloth and asks each of the 220 inhabitants, in turn, to slip his hand under the cloth, pick the first coin they touch without revealing its denomination and to answer as follows:

> If the coin is a dime, answer "Are you being held here against your will?";
> If the coin is a nickel, respond "yes";
> If the coin is a quarter, respond "no."

Estimate the proportion who wish to leave, assuming all tell the truth and he received 48 "yes" answers. Do you think the selection of a coin under this scheme would really be random?

13. An instructor of a sixth-grade class suspects that some of her students are smoking cigarettes—a practice which is against school policy. To test her conjecture, she explains the following experiment to her class and convinces them that it will be

impossible for her to tell whether any specific individual smokes or not. Each of the 32 students privately draws a marble from a well-mixed bag containing 18 red marbles, 3 green marbles, and 3 yellow marbles. After noticing the color of the marble, the student replaces it in the bag. If the student picks a red marble, the student must answer the question: "Have you smoked cigarettes in the past week?" If the student draws a green marble, the student must automatically answer "yes." The student automatically responds "no" if a yellow marble is drawn. If 10 students answer "yes", estimate the number of students in the class who have smoked cigarettes in the past week.

5. Conclusion

"...all randomized response models possess one essential feature...."

We have examined the original Warner model and two variations, the Simmons and Greenberg models. Although the Simmons and Greenberg models have certain advantages over their predecessors, all randomized response models possess the one essential feature first proposed by Warner: A randomizing device determines which of a number of questions the subject will answer without divulging this choice to the investigator. In this way, even though no information is gained from any single response, a large number of responses allows quite useful information to be easily and reliably obtained. Accordingly, the temptation to intentionally falsify answers regarding sensitive or incriminating questions, or to fail to respond altogether, is substantially reduced.

The degree of success of this method depends on many considerations, some of which are discussed in the exercises. Not the least of these is the absolute necessity of establishing rapport and trust between respondent and interviewer. In order to have the respondent answer truthfully, it is essential that the respondent be thoroughly convinced of the anonymity provided by the randomizing device— remember, the probability of selecting the sensitive question must not be so high as to incite the skepticism of the respondent. The procedure must be adequately explained and understood by the subject.

The randomized response technique has been used to measure the rates of abortion, alcohol and drug abuse, divorce, and bankruptcy, to name a few examples. Nevertheless, it should be noted that even though the randomized response technique seems to be an obvious improvement over the direct question method it is not quite the panacea one may have hoped for. As with most developments, there are costs. If precision is to be maintained, increased sample size is one of these costs. Further, the method is often more elaborate and time-consuming than direct questioning. Still, the point is to accurately estimate the proportion who respond "yes" to the sensitive question. In this respect, the reduction of answer bias establishes the randomized response technique as a very powerful tool with wide application.

Additional Exercises

14. The choice of a randomizing device is an important consideration—a device appropriate for one situation may be inappropriate for another situation. Why would a deck of cards be an unsuitable randomizing device to use to poll convention of magicians about divulging professional secrets?

15. Which of the randomized models do you feel would provide the highest degree of truthful responses? Can you think of an experiment to test your conjecture?

16. Discuss possible topics in which the randomized response technique could be used to advantage. How would you design and administer a survey to study each topic? Which model and what type of randomizing device would you employ?

17. Derive a formula for the proportion who belong to a group A with some sensitive characteristic under the following variation of a randomized response model (a tree diagram may be helpful). An urn containing 40 blue marbles and 10 white marbles is presented to a subject, who is instructed to draw a marble and then answer according to the following scheme:

   If you are a member of group A answer "yes" or "no" to the statement: "The marble I drew was blue."
   If you are not a member of group A then answer "yes" or "no" to the statement: "The marble I drew was white."

   After deriving the mathematical formulation of the experiment, solve for the proportion who belong to group A. This procedure is almost identical to which of the three models discussed in this module?

18. Accusations have been made at a large hospital that some of the doctors may be performing intentionally "needless" surgery. In order to gauge the magnitude of the problem, all 250 doctors are required to participate in a randomized response experiment.

The procedure is explained in detail so that the doctors understand that their anonymity will be protected. Each doctor tosses a pair of dice out of the view of the interviewer. If doubles are rolled, the doctor responds to the statement: "The sum of the spots on the dice is even." Otherwise, the doctor replies to the statement: "While at this hospital I have recommended or performed an operation which I knew was not necessary for the patient." If 50 "yes" answers are found, what proportion of the doctors tested perform "needless" surgery?

19. A research firm is contracted to estimate the proportion of college males in a certain region who regularly watch daytime soap operas. Because some males may be reluctant to admit to doing so, the following experiment is performed. Twenty-eight dominoes (described as ordered pairs in **Figure 5** below) are shuffled face down. The subject chooses a tile. If his selection is an odd double, i.e. (5,5) (3,3) or (1,1), then he must answer "no." If he chooses one of the tiles (6,6), (4,4) or (2,2), he must respond "yes." If he chooses any of the remaining tiles, he answers: "I usually watch one or more daytime soap operas at least once a week." A random sample of 400 college males is involved in this experiment, with 160 "yes" answers recorded. Estimate the proportion of college males in the region who watch daytime soap operas.

20. Assume the subjects in an experiment using the randomized response technique are randomly selected from some population. Let λ be the probability of a "yes" response. Then the number of "yes" answers follows a binomial distribution with parameter λ. For the politician example using the Warner model and the abortion example using the Simmons model, derive 90% and 95% confidence intervals (λ_a, λ_b) for λ. Then solve the mathematical statement of the model for π_a and π_b to find 90% and 95% confidence intervals for π.

(6,6) (5,5) (4,4) (3,3) (2,2) (1,1) (0,0)
(6,5) (5,4) (4,3) (3,2) (2,1) (1,0)
(6,4) (5,3) (4,2) (3,1) (2,0)
(6,3) (5,2) (4,1) (3,0)
(6,2) (5,1) (4,0)
(6,1) (5,0)
(6,0)

Figure 5. The 28 dominoes (zero = blank).

6. Further Readings and References

The interested reader is encouraged to consult the articles below for further details on the randomized response technique. Points discussed in these articles include applications, extensions, and generalizations of the technique, accuracy of estimates, and the investigator's responsibilities to the respondent.

Abernathy, James R., Bernard G. Greenberg, and Daniel G. Horvitz. 1970. Estimates of induced abortion in urban North Carolina. *Demography* 7:19–29.

Abul-Ela, Abdel-Latif A., Bernard G. Greenberg, and Daniel G. Horvitz. 1967. A multi-proportions randomized response model. *Journal of the American Statistical Association* 62:990–1008.

Barth, Jeffrey T., and Howard M. Sandler. 1976. Evaluation of the randomized response technique in a drinking survey. *Journal of Studies on Alcohol* 37:690–693.

Bhattacharyya, Gouri K., and Richard A. Johnson. 1977. *Statistical Concepts and Methods*. New York: Wiley.

Campbell, Cathy, and Brian L. Joiner. 1973. How to get the answer without being sure you've asked the question. *American Statistician* 27:229–231.

Frankel, Lester R. 1976. Statistics and people—the statistician's responsibilities. *Journal of the American Statistical Association* 71:9–16.

Greenberg, Bernard G., Abdel-Latif A., Abdul-Ela, Walt R. Simmons, and Daniel G. Horvitz. 1969. The unrelated question randomized response model, theoretical framework. *Journal of the American Statistical Association* 64:520–539.

Greenberg, Bernard G., Roy T. Kuebler, Jr., James R. Abernathy, and Daniel G. Horvitz. 1971. Application of randomized response technique in obtaining quantitative data. *Journal of the American Statistical Association* 66:243–250.

Horvitz, Daniel G., B.V. Shah, and Walt R. Simmons. 1967. The unrelated question randomized response model. *Proceedings of the Social Statistics Section*. Washington, D.C.: American Statistical Association.

Moors, J.J.A. 1971. Optimization of the unrelated question randomized response model. *Journal of the American Statistical Association*. 66:627–629.

Warner, Stanley L. 1965. Randomized response: A survey technique for eliminating evasive answer bias. *Journal of the American Statistical Association* 60:63–69.

Warner, Stanley L. 1971. The linear randomized response model. *Journal of the American Statistical Association* 66:884–888.

7. Answers to Exercises

1. $1/2 = (4/5)\pi + (1/5)(1 - \pi)$

   $1/2 = \pi$

2. $P(\text{extra play}) = \frac{1}{3}\frac{2}{5} + \frac{1}{3}\frac{1}{10} + \frac{1}{3}\frac{1}{4} = \frac{1}{4}$

3. $5/8 = (5/6)\pi + (1/6)(1 - \pi)$

   $11/16 = \pi$

4. If $P = 1/2$, then $P(\text{"yes"}) = (1/2)\pi + (1/2)(1 - \pi) = 1/2$. Thus we gain no information about π since it "dropped out" of the calculations.

5. $190/800 = (30/36)\hat{\pi} + (6/36)(1 - \hat{\pi})$

   $\hat{\pi} = 17/160 = 0.10625$

6. $375/1000 = (39/52)\hat{\pi} + (13/52)(1 - \hat{\pi})$

   $1/4 = \hat{\pi}$

7. $23/40 = (7/10)\hat{\pi}_s + (1 - 7/10)(1/4)$

   $5/7 = \hat{\pi}_s$

8. $65/300 = (80/100)\hat{\pi}_s + (10/100)(1/2)$

   $\hat{\pi}_s = 5/24 \doteq 0.208$

9. $45/59 = (3/4)\hat{\pi}_s + (1/4)(1/2)$

   $\hat{\pi}_s = 301/354 \doteq 0.85$

10. $\pi_u = \dfrac{\lambda_2 P_1 - \lambda_1 P_2}{P_1 - P_2}, \ P_1 \neq P_2.$

    $\pi_s = \dfrac{\lambda_1 - \lambda_2}{P_1 - P_2} + \pi_u = \dfrac{\lambda_1 - \lambda_2 + (\lambda_2 P_1 - \lambda_1 P_2)}{P_1 - P_2}$

    $= \dfrac{\lambda_1(1 - P_2) - \lambda_2(1 - P_1')}{P_1 - P_2}, \ P_1 \neq P_2.$

11. $2/5 = (4/5)\pi + 1/10$

    $3/8 = \pi$

12. $48/220 = (70/100)\hat{\pi} + 15/100$

    $\hat{\pi} = 15/154 \doteq 0.0974$. (The selection process may not be random since the coins are of different sizes.)

13. $10/32 = (18/24)\hat{\pi} + 3/24$

    $1/4 = \hat{\pi}$. Hence we estimate $(1/4)(32) = 8$ students have smoked in the past week.

17.

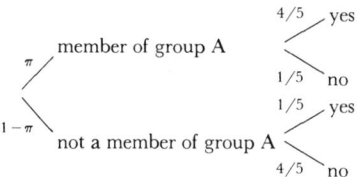

$P(\text{"yes"}) = 4/5\pi + (1/5)(1 - \pi)$. Solving for π gives $\pi = (5P(\text{"yes"}) - 1)/3$. Notice that the mathematical statement of this model is exactly the same as in the Warner model.

18. $50/250 = (30/36)\hat{\pi} + (6/36)(1/2)$

$\hat{\pi} = 7/50 = 0.14$

19. $160/400 = (22/28)\hat{\pi} + 3/28$

$\hat{\pi} = 41/110 \doteq 0.373$

20. A $(1 - \alpha)100\%$ confidence interval for the binomial parameter λ is

$$\left(\hat{\lambda} - z_{\alpha/2}\sqrt{\frac{\hat{\lambda}(1 - \hat{\lambda})}{n}}, \hat{\lambda} + z_{\alpha/2}\sqrt{\frac{\hat{\lambda}(1 - \hat{\lambda})}{n}}\right),$$

where n is the sample size ($n \geq 30$), $\hat{\lambda}$ is the observed proportion of "yes" responses, and $z_{\alpha/2}$ is the upper $\alpha/2$ point of the standard normal distribution (i.e., the area to the right of $z_{\alpha/2}$ is $\alpha/2$). The $z_{\alpha/2}$ values corresponding to the 90% and 95% confidence intervals are $z_{.05} = 1.645$ an $z_{.025} = 1.96$ respectively. For the politician example using the Warner model, $n = 1000$, $\hat{\lambda} = 300/1000 = 0.03$. Thus

$$\left(0.3 - 1.645\sqrt{\frac{(.3)(.7)}{1000}}, 0.3 + 1.645\sqrt{\frac{(.3)(.7)}{1000}}\right)$$

$$= (0.2762, 0.3238)$$

is a 90% confidence interval for λ and

$$\left(0.3 - 1.96\sqrt{\frac{(.3)(.7)}{1000}}, 0.3 + 1.96\sqrt{\frac{(.3)(.7)}{1000}}\right)$$

$$= (0.2716, 0.3284)$$

is a 95% confidence interval for λ. Since $\lambda = P\pi + (1 - P)(1 - \pi)$, we substitute the lower (λ_a) and upper (λ_b) limits of the confidence intervals to obtain confidence intervals for π. For a 90% confidence interval for π, we find the lower limit of the interval from

$$0.2762 = (3/4)\pi_a + (1/4)(1 - \pi_a).$$

Hence, $0.0524 = \pi_a$.

Similarly, since

$$0.3238 = (3/4)\pi_b + (1/4)(1 - \pi_b),$$

the upper limit is

$$0.1476 = \pi_b.$$

Consequently, a 90% confidence interval for π is $(0.0524, 0.1476)$. Following this same procedure for a 95% confidence interval for π gives the interval $(0.0432, 0.1568)$.

The abortion examples yields 90% and 95% confidence intervals for λ of $(0.03874, 0.05878)$ and $(0.03683, 0.06070)$ respectively. We then find a 90% confidence interval for π_s to be $(0.01995, 0.04857)$, and a 95% confidence interval for π_s is $(0.01721, 0.05131)$.

UMAP Module 674

Modules in Undergraduate Mathematics and its Applications

Price Elasticity of Demand: Gambling, Heroin, Marijuana, Whiskey, Prostitution, and Fish

Yves Nievergelt

Published in cooperation with the Society for Industrial and Applied Mathematics, the Mathematical Association of America, the National Council of Teachers of Mathematics, the American Mathematical Association of Two-Year Colleges, The Institute of Management Sciences, and the American Statistical Association.

INTERMODULAR DESCRIPTION SHEET:	UMAP Unit 674
TITLE:	PRICE ELASTICITY OF DEMAND: GAMBLING, HEROIN, MARIJUANA, WHISKEY, PROSTITUTION, AND FISH
AUTHOR:	Yves Nievergelt Department of Mathematics Eastern Washington University Cheney, WA 99004
MATHEMATICAL FIELD:	Calculus
APPLICATION FIELD:	Economics and public policy
TARGET AUDIENCE:	Students in the second term of calculus
ABSTRACT:	This module develops the mathematical concept of elasticity, expresses it in terms of logarithmic derivatives, and demonstrates how economists use it in evaluating social policies. The exercises range from routine applications to mathematical problems, followed by directed readings in the professional literature. Such reading assignments convince students that in real life mathematics is both useful and more complicated than in the classroom. Having realized that, students often take their calculus course more seriously.
PREREQUISITES:	The ability to differentiate composite functions that involve natural logarithms, exponentials, and rational functions, and the ability to calculate simple limits by algebraic manipulations.

© Copyright 1987 by COMAP, Inc. All rights reserved.

COMAP, 60 Lowell Street, Arlington, MA 02174 (617) 641-2600

Price Elasticity of Demand: Gambling, Heroin, Marijuana, Whiskey, Prostitution, and Fish

Yves Nievergelt
Department of Mathematics
Eastern Washington University
Cheney, WA 99004

Table of Contents

1. INTRODUCTION 1
2. THE ELASTICITY MEASURES THE RELATIVE
   RESPONSIVENESS OF A FUNCTION 2
3. THE ELASTICITY REPRESENTS A RATIO OF AREAS 5
4. DERIVATIVES MAY EASE THE CALCULATION
   OF THE ELASTICITY 7
5. A LOGARITHMIC TRANSFORMATION SIMPLIFIES
   ESTIMATIONS FROM REAL DATA 10
6. SYNOPSIS AND COMMENTS 14
   6.1 Mathematical Properties of the Elasticity 14
   6.2 Practical Applications of the Elasticity 15
7. EXERCISES 16
   7.1 First Level: Routine Exercises 16
   7.2 Second Level: Mathematical Problems 18
   7.3 Third Level: Directed Applied Reading 19
8. MODEL EXAMINATION 20
9. SOLUTIONS TO ALL THE EXERCISES 21
   9.1 First Level: Routine Exercises 21
   9.2 Second Level: Mathematical Problems 22
   9.3 Third Level: Directed Applied Reading 24
10. SOLUTIONS TO THE MODEL EXAMINATION 25
11. REFERENCES 26

Modules and Monographs in Undergraduate
Mathematics and Its Applications (UMAP) Project

The goal of UMAP was to develop, through a community of users and developers, a system of instructional modules in undergraduate mathematics and its applications to be used to supplement existing courses and from which complete courses may eventually be built.

The Project was guided by a National Advisory Board of mathematicians, scientists, and educators. UMAP was funded by a grant from the National Science Foundation and is now supported by the Consortium for Mathematics and Its Applications (COMAP), Inc. a nonprofit corporation engaged in research and development in mathematics education.

COMAP Staff

Solomon A. Garfunkel	Executive Director, COMAP
Laurie W. Aragon	Business Development Manager
Philip A. McGaw	Production Manager
Nancy Hawley	Copy Editor
Annemarie S. Morgan	Administrative Assistant

UMAP Advisory Board

Steven J. Brams	New York University
Llayron Clarkson	Texas Southern University
Donald A. Larson	SUNY at Buffalo
R. Duncan Luce	Harvard University
Frederick Mosteller	Harvard University
George M. Miller	Nassau Community College
Walter Sears	University of Michigan Press
Arnold A. Strassenburg	SUNY at Stony Brook
Alfred B. Willcox	Mathematical Association of America

This module was developed and tested in the preparatory mathematics course of the Executive Master of Business Administration (EMBA) Program, at the University of Washington in Seattle, WA.

1. Introduction

Econometric studies typically encompass entire economic sectors, for instance the traffic of heroin in Detroit, or the gambling industry in Nevada. To evaluate related legal and fiscal policies, such studies often focus on the effect of a price change upon the volume of sales and total revenues.

"... an increase in the price of heroin may cause a rise in urban crime...."

As an introductory example (examined in greater detail further below), consider the analysis of the heroin market in Detroit by Silverman and his collaborators [Brown and Silverman 1973; Silverman and Spruill 1977]. Their results indicate that an increase in the price of heroin may cause a rise in urban crime, because consumers need more money to sustain their addiction. Harassing heroin dealers may then increase the crime rate, because such harassment reduces the supply of heroin and hence raises its price. This conclusion contrasts with that reached by Nisbet and Vakil on the market for marijuana at the University of California in Los Angeles [Nisbet and Vakil 1972]. They found that an increase in the price of marijuana tends to reduce the total amount of money spent on the drug, which may lessen the rate of related crime. In this case, prosecuting dealers decreases both the consumption of marijuana and the crimes to finance it.

"... an increase in the price of marijuana may lessen the rate of related crime...."

These two examples—heroin and marijuana—show that to be efficient, law enforcement strategies may have to differ from one market to another. Yet these examples do not imply that law enforcement strategies should depend upon considerations of price only. Indeed, economists appear to disagree among themselves as to whether standard market analyses apply to illegal drug markets [Silverman and Spruill 1977; White and Luksetich 1983], and thence they also seem to disagree on the potential consequences of legalizing and taxing such drugs as marijuana [Kleiman 1986; Zeese 1986]. Of course the present discussion does not advocate any practical solution, but focuses strictly on the mathematics most frequently used by economists in their market studies.

To evaluate social policies in the manner outlined above, economists need a measure of market responsiveness that is independent of units and market size, for instance to allow for comparisons between markets. To this end they first model the market for a commodity by a *demand function*, $f: p \mapsto q = f(p)$, which relates the unit price p, say in dollars per item, to the quantity $q = f(p)$ sold in a given period, say in items per month. Economists call the graph of a demand function a *demand curve*. For example, **Figure 1** displays a demand curve for heroin in Detroit; the horizontal scale represents the unit price p in dollars per gram of 22% pure heroin, while the

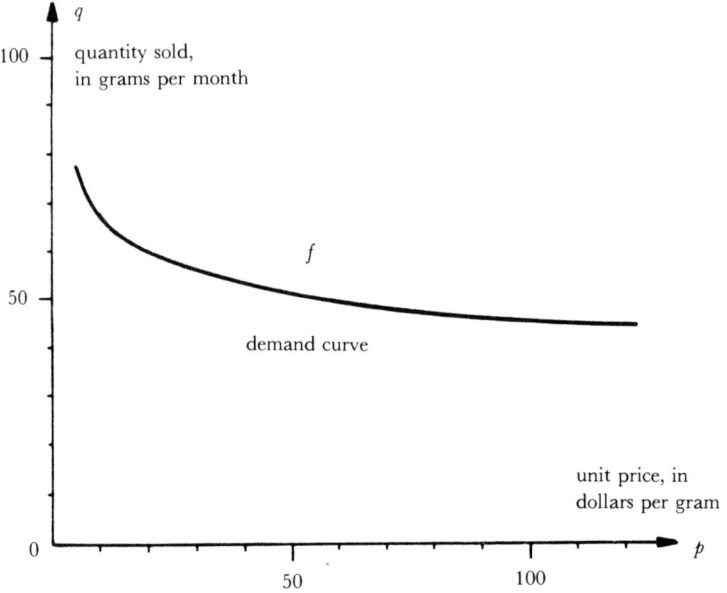

Figure 1. The demand for heroin in Detroit.

vertical scale shows the quantity sold q in grams per month per consumer [Silverman et al., op. cit.].

To determine the responsiveness of the market to a price change, economists then calculate the "price elasticity of demand," a quantity defined in terms of logarithmic derivatives of the demand function. To clarify these concepts, this module defines them in mathematical terms, explains how economists use them, and how they may help the decisions of public officials. The exercises also demonstrate other uses of elasticity: for instance, income elasticity of demand and survival elasticity of salmon populations.

2. The Elasticity Measures the Relative Responsiveness of a Function

Such practical problems as those just described involve some function f—for instance, a demand function—and their solution requires a measure of responsiveness of that function. This measure relies on the following notion of "relative change."

DEFINITION 1. *The relative change from a number x_1 (different from zero) to another number x_2 is the ratio $\Delta\% x = (x_2 - x_1)/x_1$. Thus the relative change, $\Delta\% x$, expresses the difference $x_2 - x_1$ as a fraction of the initial value x_1.*

Example 1. Suppose that $x_1 = 3$ and $x_2 = 2.4$, for instance. In this particular case (with the symbol "$\rightleftharpoons$" meaning "corresponding to"), the relative change becomes

$$\Delta\% x = \frac{x_2 - x_1}{x_1} = \frac{2.4 - 3}{3} = \frac{-0.6}{3}$$

$$= -0.2 = -\frac{20}{100} \rightleftharpoons -20\%.$$

Thus the difference, -0.6, represents a drop of 20% relatively to the initial value, 3.

A change from x_1 to x_2 also induces a change $y_2 - y_1$ in the values of the function f under consideration, from $y_1 = f(x_1)$ to $y_2 = f(x_2)$, to which corresponds the relative change $\Delta\% y = (y_2 - y_1)/y_1$.

Example 2. Suppose that $f(x) = 1/x$, for instance, and consider the change from $x_1 = 3$ to $x_2 = 2.4$. Then $y_1 = f(3) = 1/3$ and $y_2 = f(2.4) = 1/2.4$. Consequently,

$$\Delta\% y = \frac{y_2 - y_1}{y_1} = \frac{1/2.4 - 1/3}{1/3} = \frac{1/12}{1/3} = \frac{1}{4} = 0.25 \rightleftharpoons 25\%.$$

This means that the difference $y_2 - y_1$ represents an increase of 25% with respect to y_1.

This concept of relative change leads to a convenient expression for a measure of responsiveness called "arc elasticity."

DEFINITION 2. *The arc elasticity of a function f, from x_1 to x_2, is the following ratio, usually denoted by η (the Greek letter "eta"),*

$$\eta = \frac{\Delta\% y}{\Delta\% x} = \frac{(y_2 - y_1)/y_1}{(x_2 - x_1)/x_1} = \frac{(f(x_2) - f(x_1))/f(x_1)}{(x_2 - x_1)/x_1}.$$

The arc elasticity η measures the ratio of the relative change $\Delta\% y$ to the relative change $\Delta\% x$, and thus it reflects the responsiveness of f to changes in x.

Example 3. Suppose that $f(x) = 1/x$ and that x changes from $x_1 = 3$ to $x_2 = 2.4$, as before. Then

$$\eta = \frac{\Delta\%y}{\Delta\%x} = \frac{(1/2.4 - 1/3)(1/3)}{(2.4 - 3)/3} = \frac{1/4}{-1/5}$$
$$= -1.25 \rightleftharpoons -125\%.$$

This figure means that the relative change in y ($1/4 \rightleftharpoons 25\%$) equals -125% of the relative change in x ($-1/5 \rightleftharpoons -20\%$).

While the arc elasticity requires two points x_1 and x_2, some situations involve only one point x_1, especially in the computations for fitting demand curves to real data. In those cases another measure of responsiveness, simply called "elasticity," appears useful.

DEFINITION 3. *The elasticity of a function f at a point x_1 is the following limit (if it exists), usually denoted by $\eta_f(x_1)$,*

$$\eta_f(x_1) = \lim_{x_2 \to x_1} \frac{\Delta\%y}{\Delta\%x} = \lim_{x_2 \to x_1} \frac{(y_2 - y_1)/y_1}{(x_2 - x_1)/x_1}$$
$$= \lim_{x_2 \to x_1} \frac{(f(x_2) - f(x_1))/f(x_1)}{(x_2 - x_1)/x_1}.$$

This definition means that as x_2 tends to x_1, the arc elasticity from x_1 to x_2 approaches the elasticity of f at the single point x_1. In this sense, $\eta_f(x_1)$ characterizes the responsiveness of f in the vicinity of x_1.

Example 4. Suppose that $f(x) = 1/x$, as in the previous examples. Then

$$\eta_f(x_1) = \lim_{x_2 \to x_1} \frac{\Delta\%y}{\Delta\%x} = \lim_{x_2 \to x_1} \frac{(y_2 - y_1)/y_1}{(x_2 - x_1)/x_1}$$
$$= \lim_{x_2 \to x_1} \frac{(1/x_2 - 1/x_1)/(1/x_1)}{(x_2 - x_1)/x_1}.$$

Multiplying the numerator and the denominator by $x_1 x_2$ simplifies this expression to

$$\eta_f(x_1) = \lim_{x_2 \to x_1} \frac{(x_1/x_2 - x_1/x_1)x_1 x_2}{((x_2 - x_1)/x_1)x_1 x_2}$$
$$= \lim_{x_2 \to x_1} \frac{(x_1 - x_2)x_1}{(x_2 - x_1)x_2} = \lim_{x_2 \to x_1} -\frac{x_1}{x_2} = -1.$$

Hence $\eta_f(x_1) = -1$; the elasticity of f equals -1 at every point $x_1 \neq 0$.

3. The Elasticity Represents a Ratio of Areas

Figure 2 shows a geometric interpretation of the elasticity. Consider the rectangle with vertices at $(0,0)$, $(x_1, 0)$, (x_1, y_1), and $(0, y_1)$, the area of which equals $R(x_1) = |x_1 y_1|$. (The absolute value takes care of possible negative x_1 or y_1.) Increasing x from x_1 to x_2 subtracts the horizontal rectangle labeled "LONG" with length $|x_1|$ and height $|y_2 - y_1|$, and adds the vertical rectangle labeled "HIGH" with length $|x_2 - x_1|$ for height $|y_2|$. Thus the area of the initial rectangle decreases by the area $|(y_2 - y_1)x_1|$ and increases by the area $|(x_2 - x_1)y_2|$; the ratio of these two areas is then

$$\frac{\text{Decrease}}{\text{Increase}} = \frac{\text{LONG}}{\text{HIGH}} = \frac{|(y_2 - y_1)x_1|}{|(x_2 - x_1)y_2|} = \frac{|(y_2 - y_1)/y_2|}{|(x_2 - x_1)/x_1|}.$$

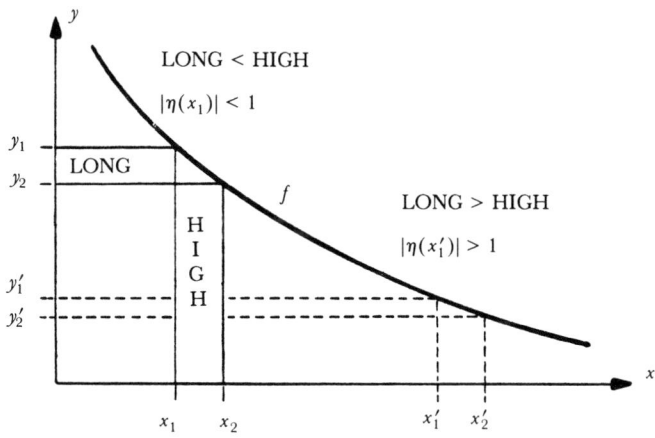

Figure 2. As x_1 increases to x_2, the area $R(x_1)$ changes to $R(x_2)$, decreasing by the "LONG" area and increasing by the "HIGH" area. The elasticity $\eta_f(x_1)$ measures the limit of the ratio LONG/HIGH.

As x_2 tends to x_1, y_2 tends to y_1 by continuity, and y_2/y_1 approaches 1. Hence

$$\lim_{x_2 \to x_1} \frac{\text{Decrease}}{\text{Increase}} = \lim_{x_2 \to x_1} \frac{|(y_2 - y_1)/y_2|}{|(x_2 - x_1)/x_1|} \frac{|y_2|}{|y_1|}$$

$$= \lim_{x_2 \to x_1} \frac{|(y_2 - y_1)/y_1|}{|(x_2 - x_1)/x_1|} = |\eta_f(x_1)|.$$

Thus the absolute value of the elasticity is the limit of the ratio of the decrease to the increase in the area of the rectangle with vertices at $(0, 0)$, $(x_1, 0)$, (x_1, y_1), and $(0, y_1)$.

The rectangle and the elasticity also have economic interpretation when the function represents a demand curve. In that context, x stands for the price per unit and y for the quantity sold per period. Since revenues equal price times quantity, then the area of the rectangle, $R(x) = xy$, equals the revenue generated by the sale of y items at unit price x each. Economists distinguish three situations:

- If $-1 < \eta(x) < 0$, then $|\eta(x)| < 1$ and the increase in area surpasses its decrease; in economic words, the revenues rise as the price increases. In this situation the demand is said to be *inelastic* at x. With an inelastic demand, the quantity does not react much to a change in price so that revenues rise if prices go up.
- If $\eta(x) = -1$, then $|\eta(x)| = 1$ so that the decrease and the increase in area cancel each other, hence the revenues remain approximately the same. Economists then say that the demand is *unit elastic*. With a unit elastic demand, the quantity sold reacts so as to offset a price change exactly, which keeps the revenues unchanged.
- If $\eta(x) < -1$, then $|\eta(x)| > 1$, and in this case the increase in area does not suffice to offset its decrease; hence the revenues decline as the price rises. The demand is then said to be *elastic* at x. With an elastic demand, the quantity sold reacts to a price change by a large amount so that revenues drop if prices rise.

Application 1. Hogarty and Elzinga modeled the demand for beer with a function of the type $f(x) = 1/x$, with the price x expressed in dollars per can of beer, and the quantity sold $f(x)$ in cans per day per adult [Hogarty and Elzinga 1972]. As shown above for this function, $\eta_f(x) = -1$ at every price x (except at zero), and thus the revenues remain constant when the price changes; hence the demand for beer is unit elastic. The elasticity also provides an approximation of the relative change in consumption due to a relative change in the price: from the definition of $\eta_f(x_1)$ as a limit,

$$\eta_f(x_1) = \lim_{x_2 \to x_1} \frac{\Delta\% y}{\Delta\% x},$$

it follows that for "small" changes $|x_2 - x_1|$,

$$\eta_f(x_1) \approx \frac{\Delta\% y}{\Delta\% x}$$

(with the symbol "$\approx$" meaning "approximately"); therefore

$$\Delta\% y \approx \eta_f(x_1) \cdot \Delta\% x.$$

For example, if the price of beer rises by, say, 2%, then $\Delta\% x \rightleftharpoons 2\%$ and hence the consumption drops by approximately $\Delta\% y \approx \eta_f(x_1) \cdot \Delta\% x = (-1) \cdot (2\%) \rightleftharpoons -2\%$.

4. Derivatives May Ease the Calculation of the Elasticity

As the above example illustrates, using limits to calculate the elasticity of such an elementary function as that given by $f(x) = 1/x$ involves cumbersome algebraic manipulations. Fortunately, a simple relationship between the elasticity $\eta_f(x)$ and the derivative $f'(x)$ eases this calculation considerably. To arrive at that relationship, recall that

$$\eta_f(x) = \lim_{x_2 \to x} \frac{(f(x_2) - f(x))/f(x)}{(x_2 - x)/x}.$$

Under the assumption that the derivative of f exists (in addition to the condition that $f(x) \neq 0$ to avoid a division by zero), rearranging terms gives

$$\eta_f(x) = \lim_{x_2 \to x} \frac{x \cdot (f(x_2) - f(x))}{f(x) \cdot (x_2 - x)}$$

$$= \frac{x}{f(x)} \cdot \lim_{x_2 \to x} \frac{f(x_2) - f(x)}{x_2 - x} = \frac{x}{f(x)} \cdot f'(x).$$

Consequently,

$$\eta_f(x) = \frac{x}{f(x)} \cdot f'(x) = \frac{x}{y} \cdot \frac{dy}{dx}.$$

Example 5. In the previous example, with $f(x) = 1/x$, the derivative is $f'(x) = -1/x^2$. Hence the above formula becomes

$$\eta_f(x) = \frac{x}{f(x)} \cdot f'(x) = \frac{x}{1/x} \cdot \frac{-1}{x^2} = \frac{x^2}{1} \cdot \frac{-1}{x^2} = -1.$$

Therefore $\eta_f(x) = -1$ for every x (except at zero, where f is not defined), which corroborates the same result found earlier with limits.

Besides simplifying computations, derivatives also provide a second way to link the price elasticity of demand to the sales revenues. With $f(x)$ denoting the quantity sold at price x, the revenues obey the formula $R(x) = x \cdot f(x)$, and the product rule for derivatives yields $R'(x) = 1 \cdot f(x) + x \cdot f'(x)$. Factoring $f(x)$ gives

$$R'(x) = \left(1 + \frac{x}{f(x)} \cdot f'(x)\right) \cdot f'(x) = \left(1 + \eta_f(x)\right) \cdot f(x).$$

Since the quantity sold, $f(x)$, cannot be negative, then the derivative $R'(x)$ has the same sign as the factor $(1 + \eta_f(x))$, according to whether

- $\eta_f(x) < -1$, in which case $R'(x) < 0$ (the demand is *elastic*), or whether
- $\eta_f(x) > -1$, in which case $R'(x) > 0$ (the demand is *inelastic*).

In particular, if the revenues reach a maximum at some price level, say x^*, then $R'(x^*) = 0$ and therefore $\eta_f(x^*) = -1$. **Figure 3** illustrates these situations.

Application 2. Wales described the demand for distilled spirits with "affine" functions of the following type [Wales 1968]:

$$q = f(p) = -0.00375p + 7.87,$$

with p standing for the retail price of liquor, in dollars per case (of twelve fifths per case), and q representing the average number of cases of distilled spirits purchased annually by individual consumers. Since $f'(p) = -0.00375$, then

$$\eta_f(p) = \frac{p}{f(p)} \cdot f'(p) = \frac{-0.00375p}{-0.00375p + 7.87}.$$

Observe that in this example the elasticity varies with the price, whereas it was constant, -1, in the previous example. For instance,

Price Elasticity of Demand   163

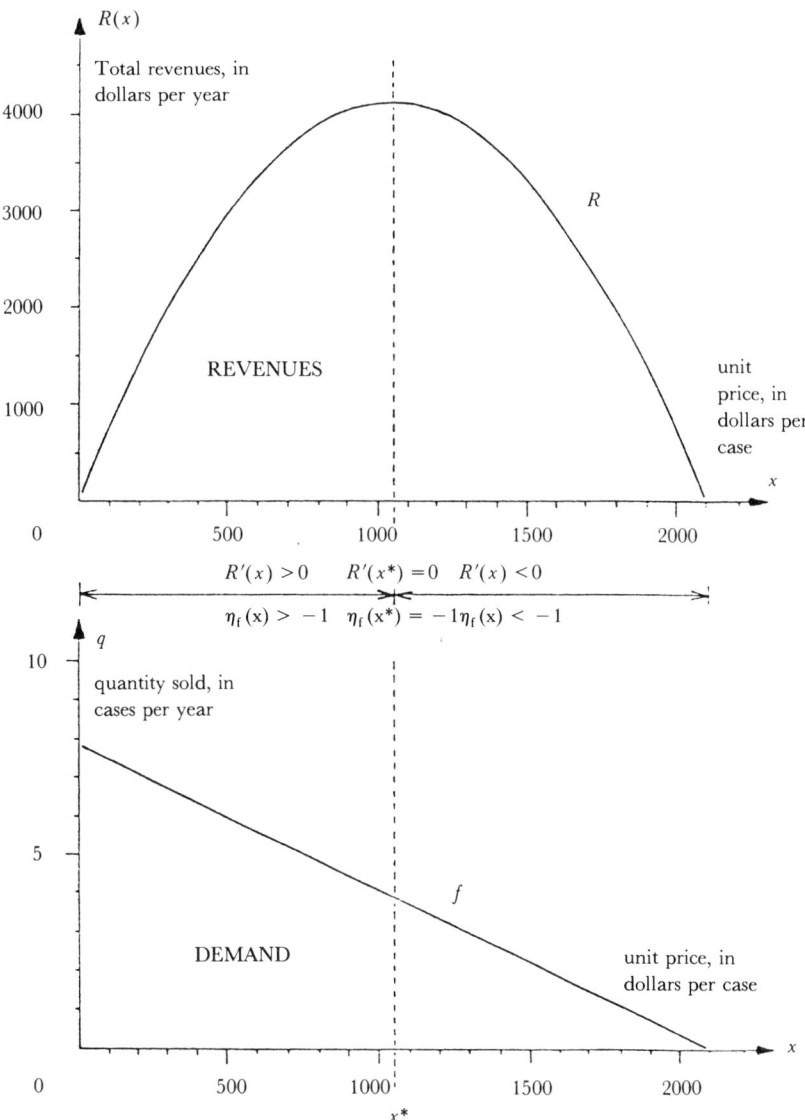

Figure 3. As the price increases, the revenues rise where the demand is inelastic, and the revenues fall where the demand is elastic.

if liquor costs $58.20 per case ($4.85 per fifth, a typical figure for Seagram 7 Crown in 1962 [Simon 1966]), then

$$\eta_f(58.20) = \frac{-0.00375 \times 58.20}{-0.00375 \times 58.20 + 7.87} \approx -0.027.$$

Since $-1 < \eta_f(58.20) < 0$, then $R'(58.20) > 0$, which means that the revenues would rise if the price increased. Furthermore, the figure $\eta_f(58.20) = -0.027$ means that a relative price increase of, say, 20% (from \$58.20 to \$69.84) would lead to a decline in consumption by about $-0.027 \times 0.20 = -0.0054$, approximately half a percent. (**Figure 3** displays the demand function, f, and the revenue function, R, that pertain to this application. Keep in mind, however, that **Figure 3** models the actual demand for distilled spirits in the vicinity of $p = 58.20$ only.)

5. A Logarithmic Transformation Simplifies Estimations from Real Data

The expression of the elasticity η in terms of derivatives, $\eta_f(x) = xf(x)/f'(x)$, reveals a relation between the elasticity and the natural logarithm of the function f. Since the derivative of the natural logarithm is $\ln'(y) = 1/y$, applying the chain rule to the composite function $\ln(f(x))$ shows that

$$\frac{d}{dx}\ln(f(x)) = \ln'(f(x)) \cdot f'(x) = \frac{1}{f(x)} \cdot f'(x),$$

whence

$$\frac{d\ln(f(x))/dx}{d\ln(x)/dx} = \frac{f'(x)/f(x)}{1/x} = \frac{x}{f(x)} \cdot f'(x) = \eta_f(x).$$

Therefore

$$\eta_f(x) = \frac{d\ln(f(x))}{d\ln(x)}.$$

This formula means that the elasticity represents the rate of change of $\ln(f(x))$ with respect to $\ln(x)$; in other terms, $\eta_f(x)$ is the slope of the graph of $\ln(f(x))$ versus $\ln(x)$ (yet the elasticity is *not* the slope of f).

Example 6. Suppose that $f(x) = 1/x$, as in the previous examples, and recall that one property of the logarithm states that $\ln(1/x) = -\ln(x)$. Consequently,

$$\eta_f(x) = \frac{d\ln(f(x))}{d\ln(x)} = \frac{-d\ln(x)}{d\ln(x)} = \frac{-1/x}{1/x} = -1.$$

With the change of variables $Y = \ln(y)$ and $X = \ln(x)$, the formula becomes

$$\eta_f(x) = \frac{dY}{dX}.$$

In practice, the logarithmic transformation of (x, y) into (X, Y) facilitates the estimation of the elasticity η by avoiding further limits and derivatives, as the following application demonstrates.

Application 3. Suits collected the following data on the demand for services of sports-betting parlors in Nevada [Suits 1979]. The total volume of bets, $q = f(p)$ (called the *handle*), expressed in millions of dollars per quarter of a year, depends upon the price, p (the *take-out rate*), in dollars per dollar wagered, which operators withhold on every bet. The demand curve f passes through the two points shown in **Figure 4(a)**, $(p_1, q_1) = (0.10, 4.957)$ and $(p_2, q_2) = (0.18, 1.386)$. These two data points alone do not suffice to specify the entire demand curve, because there exist many curves that pass through both points. At this stage, economists choose the curve that best fits both their data and their purpose. With the purpose of estimating the price elasticity of demand, they often find curves with constant elasticity convenient, and they proceed as follows. Setting $P = \ln(p)$ and $Q = \ln(q)$ transforms the data into two new points, shown in **Figure 4(b)**,

$$(P_1, Q_1) = (\ln(p_1), \ln(q_1)) = (\ln(0.10), \ln(4.957))$$
$$= (-2.303, 1.601),$$
$$(P_2, Q_2) = (\ln(p_2), \ln(q_2)) = (\ln(0.18), \ln(1.386))$$
$$= (-1.715, 0.3264).$$

The straight line through the two new points provides a simple way to estimate the price elasticity of demand for this market. The standard two-point formula gives the line's equation:

$$Q - 1.601 = \frac{0.3264 - 1.601}{-1.715 - (-2.303)} \cdot (P - (-2.303))$$
$$= -2.17P - 3.39.$$

Figure 4(b) displays this straight line. Rearranging terms yields $Q = -2.17P - 3.39$, hence

$$\eta = \frac{dQ}{dP} = -2.17.$$

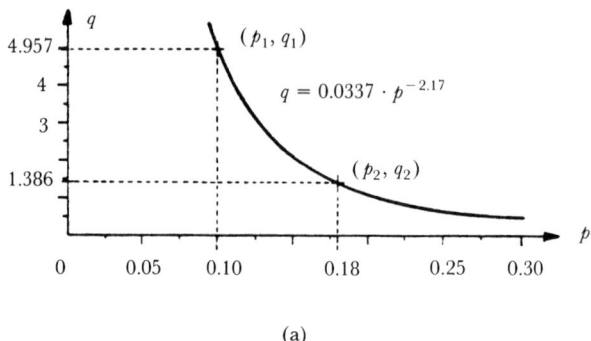

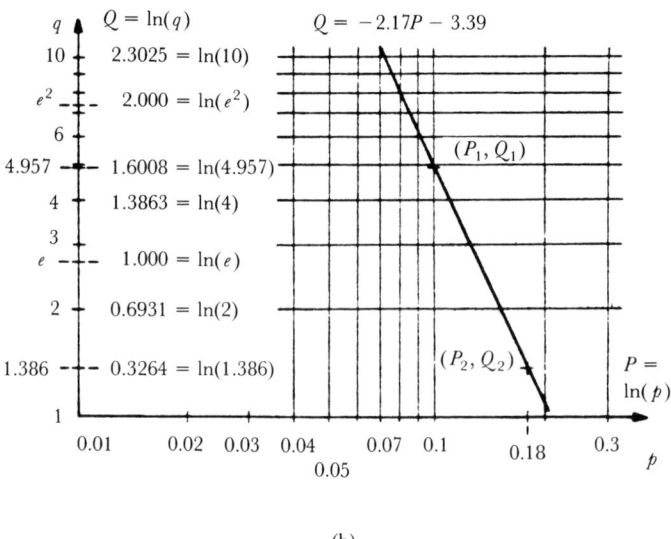

Figure 4. (a) The two data points and the final fitted demand curve with constant elasticity (but not constant slope). (b) The transformed data points and the fitted straight line, which corresponds to the demand curve in (a).

Since $\eta < -1$, we have $|\eta| > 1$ and the demand is elastic, which means that an increase in the take-out rate would decrease the handle. As discussed by Suits, legislators may find this information useful when considering tax increases to raise the state's revenues.

The fitted equation, $Q = -2.17P - 3.39$, also gives an estimate of the demand function f. Since $P = \ln(p)$ and $Q = \ln(q)$, then

$p = e^P$ and $q = e^Q$. Hence

$$f(p) = q = e^Q = e^{-2.17P - 3.39} = (e^P)^{-2.17} \cdot e^{-3.39}$$
$$= 0.0337 \cdot p^{-2.17}.$$

This example explains why such demand equations as $q = 0.0337 \cdot p^{-2.17}$ appear so frequently in practice: the functional form $q = C \cdot p^\eta$ yields the elasticity η immediately.

Application 4. Brown and Silverman showed that the price elasticity of demand for heroin in Detroit was about -0.17 [Brown and Silverman 1973]. They arrived at such values by computing logarithms $P = \ln(p)$ and $Q = \ln(q)$ for each data point, and then fitting a linear equation of the form $Q = -0.17P + 4.61$ to the new data. Reverting to the unit price p (in dollars per gram) and to the quantity q (in grams per month) gives $\ln(q) = 0.17 \cdot \ln(p) + 4.61$, and exponentiating both sides yields $q = 100 \cdot p^{-0.17}$, whence $\eta = -0.17$. (This demand curve appears in **Figure 1**, as mentioned in the introduction.) Because of this *inelastic* demand ($|\eta| = 0.17 < 1$), harassing heroin dealers may raise the crime rate: Such harassment reduces the supply of the drug, which increases prices and therefore generates higher revenues, some of which come from crimes. Quite differently, harassing heroin buyers lowers the demand and hence lowers prices, which in turn affects the dealers' revenues and may reduce the crime rate.

From this perspective the market for heroin in Detroit contrasts with that for marijuana at UCLA. Nisbet and Vakil modeled that market with a equation of the type $Q = -1.013P + 2.73$, which corresponds to the demand equation $q = 15.4 \cdot p^{-1.013}$ and reveals a slight elasticity ($|\eta| = 1.013 > 1$) [Nisbet and Vakil 1972]. In that case harassing marijuana dealers tends to decrease both consumption and related crime.

Question. Consider the following quote from [White and Luksetich 1983].

> Harassment of the buyers of prostitutes' services has been very effective, much more effective than harassment of the sellers has been.

For the purpose of this question, assume that, by definition, "effectiveness" measures the reduction in the prostitutes' total revenue, or, equivalently, the reduction in the total amount spent by their customers. Question: Is the demand for prostitutes' services elastic or inelastic? (See the solution to **exercise 14**.)

6. Synopsis and Comments

6.1 Mathematical Properties of the Elasticity

Intuitively, the elasticity measures the relative sensitivity of a function.

Specifically, the concept of elasticity relies first on the notion of the *relative change* from a number x_1 to another number x_2, defined as $\Delta\% x = (x_2 - x_1)/x_1$, provided $x \neq 0$. The relative change, $(x_2 - x_1)/x_1$, expresses the difference $x_2 - x_1$ as a fraction of the initial value x_1. This notion also applies to the values $y_1 = f(x_1)$ and $y_2 = f(x_2)$ of a function f, in which case it gives the relative change in y, $\Delta\% y = (y_2 - y_1)/y_1$, provided $y_1 \neq 0$.

The quotient η of the two relative changes, $\eta = (\Delta\% y)/(\Delta\% x)$, is called the *arc elasticity* of f from x_1 to x_2,

$$\eta = \frac{\Delta\% y}{\Delta\% x} = \frac{(y_2 - y_1)/y_1}{(x_2 - x_1)/x_1} = \frac{(f(x_2) - f(x_1))/f(x_1)}{(x_2 - x_1)/x_1}.$$

The arc elasticity represents a relative analog to the average rate of change,

$$\frac{f(x_2) - f(x_1)}{x_2 - x_1}.$$

As x_2 tends to x_1, the arc elasticity η may converge to a limit, denoted by $\eta_f(x_1)$ and called the *elasticity* of f at x_1, if the limit exists,

$$\eta_f(x_1) = \lim_{x_2 \to x_1} \frac{(f(x_2) - f(x_1))/f(x_1)}{(x_2 - x_1)/x_1}.$$

The elasticity measures the relative responsiveness of f to changes in the vicinity of x_1; it represents a relative analog to (but not the same notion as) the derivative,

$$f'(x_1) = \lim_{x_2 \to x_1} \frac{f(x_2) - f(x_1)}{x_2 - x_1}.$$

The elasticity and the derivative stand for analogous but distinct concepts. They are related by the following formula, which makes possible the use of calculus for computing elasticities,

$$\eta_f(x) = \frac{x}{f(x)} \cdot f'(x) = \frac{x}{y} \cdot \frac{dy}{dx}.$$

Still another formula relates the elasticity to logarithmic derivatives. With the change of variables $X = \ln(x)$ and $Y = \ln(y) = \ln(f(x))$, the elasticity takes the form

$$\eta_f(x) = \frac{d \ln(f(x))}{d \ln(x)} = \frac{dY}{dX}.$$

This means that the elasticity $\eta_f(x)$ represents the slope of the graph of $\ln(f(x))$ versus $\ln(x)$ at the point $\ln(x)$ (but not the slope of f). Therefore a function defined by such a formula as $f(x) = C \cdot x^\eta$ has a constant elasticity (but not necessarily a constant slope), because the change of variables $Y = \ln(f(x)) = \eta \cdot \ln(x) + \ln(C) = \eta \cdot X + B$ shows that $dY/dX = \eta$ remains constant.

6.2 Practical Applications of the Elasticity

In market analyses and econometric studies, the function f often plays the role of a *demand function*, which expresses the quantity $q = f(p)$ sold at the unit price p in a given period. Economists call its elasticity, η_f, the *price elasticity of demand*. This elasticity measures the relative sensitivity of the market and thus does not depend on units or market size; for this reason, it allows for comparisons between different markets. The formula in terms of logarithmic derivatives facilitates fitting to the data a demand curve with constant elasticity and hence estimating the price elasticity of demand.

The price elasticity of demand usually takes on negative values only. Indeed, $\eta_f(p) = p \cdot f'(p)/f(p)$, and in this expression both the price p and the quantity $f(p)$ are positive, but the slope of the demand curve $f'(p)$ is usually negative. (The need for the adverb "usually" arises from a minor disagreement among economists. For example, economists do not seem to agree among themselves whether the demand for potatoes in Ireland sloped upward or downward in the years 1845–1849 [Kohli 1986].) In any event, economists say that the demand is *elastic* if $-\infty < \eta_f(p) < -1$, in which case the revenues fall if the price rises. If $\eta_f(p) = -1$, then the demand is *unit elastic*; in this situation, the revenues may have a maximum or they may stay constant, depending upon the values of the elasticity on both sides of p. If $-1 < \eta_f(p) < 0$, then economists say that the demand is *inelastic*, and in this case the revenues rise if the price rises. The value of the price elasticity of demand may help economists assess the potential impacts of various law enforcement policies or marketing methods.

The concept of elasticity also applies to other variables. For instance, suppose that a function $g: x \mapsto q = g(x)$ gives the quantity demanded by the market in terms of the consumers' income, x, instead of the unit price; then the *income elasticity of demand*, η_g, measures the relative responsiveness of the market to a change in the consumers' income. (See **exercise 11** for an example.)

In such a different area as fishery, for example, the concept of elasticity may also measure the relative sensitivity of a salmon population to changes in the probabilities of survival between stages. Then a comparison of these elasticities may reveal which resource management strategies appear more reliable, as mentioned in **exercise 15**.

Nevertheless, all these elasticities are mathematically identical,

$$\eta_f(x) = \frac{x}{f(x)} \cdot f'(x) = \frac{x}{y} \cdot \frac{dy}{dx}.$$

7. Exercises

7.1 First Level: Routine Exercises

1. Silverman and Spruill related the retail price of heroin to the crime rate in Detroit by an equation of the type

$$c = 3351 \cdot p^{0.2870},$$

with c representing the number of property crimes reported monthly in Detroit, and p denoting the price of heroin in that city, in dollars per gram [Silverman and Spruill 1977].

1.1 Derive a linear equation between $\ln(c)$ and $\ln(p)$.
1.2 Find the elasticity of c with respect to p.
1.3 By approximately what percentage does the rate of property crimes c rise if the price of heroin p rises by, say, 5%?

2. The above authors also related the price of heroin p to the number C of such personal crimes as murders and rapes reported monthly in Detroit, by an equation of the form

$$C = 207.8 \cdot p^{0.3494}.$$

2.1 Find the elasticity of C with respect to p.
2.2 By approximately what percentage does the rate of personal crimes C rise if the price of heroin p increases by, say, 5%?
2.3 At an average retail price of $57.51 per gram of heroin, which rate is larger: that of property crimes, c (as in **exercise 1**), or that of personal crimes, C?

Price Elasticity of Demand 171

3. Suits collected the following two data points for off-track horse betting in Nevada [Suits 1979]. Let p denote the unit price (take-out rate), in dollars per dollar wagered, and let q stand for the total amount wagered per quarter of a year (handle), in millions of dollars. Then $(p_1, q_1) = (0.16, 9.556)$ and $(p_2, q_2) = (0.236, 5.055)$.

   3.1 Perform the transformation $P = \ln(p)$ and $Q = \ln(q)$ on the given data to obtain two new points, (P_1, Q_1) and (P_2, Q_2).
   3.2 Determine the equation $Q = mP + b$ of the straight line through (P_1, Q_1) and (P_2, Q_2). In other words, calculate m and b.
   3.3 By taking the exponential of each side of $Q = mP + b$, compute the two numbers C and η such that the demand curve given by $q = C \cdot p^\eta$ pass through the two data points (p_1, q_1) and (p_2, q_2).
   3.4 Find the elasticity of the demand function found in 3.3.
   3.5 By approximately what percentage does the quarterly handle q decrease if the take-out rate p rises by, say, 20%?

4. Nisbet and Vakil also fitted a linear demand equation of the following sort to the market for marijuana among UCLA students:

$$q = -0.225p + 3.74,$$

in which p denotes the price of marijuana in dollars per ounce, and q represents the average quantity purchased monthly in ounces per consumer [Nisbet and Vakil 1972].

   4.1 Express the price elasticity of demand, η, by a formula in terms of p.
   4.2 Compute the elasticity corresponding to the then-prevailing price of $10 per ounce.
   4.3 By approximately what percentage should the price of marijuana increase for the consumption to decrease by, say, 5% (from the prevailing level of 1.49 ounces per month down to 1.42)?
   4.4 Determine the price at which the demand for marijuana is unit elastic. In other words, find the price p^* such that $\eta(p^*) = -1$.

5. Simon estimated the price elasticity of demand for Seagram 7 Crown blended whiskey about at -0.79 [Simon 1966]. According to his study, per capita liquor consumption averaged two fifths per adult per year in 1961, and Seagram cost $4.85 per fifth then.

5.1 With the data provided above, determine two numbers M and B, such that the demand curve given by $\ln(q) = M \cdot \ln(p) + B$, pass through the point $(p_1, q_1) = (4.85, 2)$ with elasticity equal to -0.79. Then express the demand curve in the form $q = C \cdot p^\eta$.

5.2 Suppose that the price of liquor rises from \$4.85 to \$5.00. What is the relative price change? By approximately what percentage does per capita consumption decrease? Does the total amount of money spent on liquor increase or decrease?

5.3 Find the slope-intercept equation, $q = mp + b$, of the straight line that passes through the point $(4.85, 2)$ with elasticity -0.79 at that point. (Finding M and B may involve a system of two linear equations.) This question demonstrates that more than one curve may fit a set of data.

7.2 Second Level: Mathematical Problems

6. Let f be an "affine" function, of the type $f(x) = mx + b$, with slope m and verticle intercept b. Prove that $\eta_f(x)$ equals the arc elasticity of f from x to every second point w, regardless of where w lies. Then provide a counterexample showing that equality may fail if f is not affine.

7. Suppose that the elasticity $\eta_g(x)$ of some function g always equals the arc elasticity of g from x to every second point w, regardless of w. Prove that g must be an affine function (with a straight line as its graph).

8. Let q be an "affine" function, given by $q(p) = mp + b$, with negative slope, $m < 0$, and positive vertical intercept, $b > 0$. Prove that

8.1 $-1 < \eta_q(p) < 0$ for every p between 0 and $-b/(2m)$;

8.2 $\eta_q(p^*) = -1$ for $p^* = -b/(2m)$;

8.3 $-\infty < \eta_q(p) < -1$ for every p between $-b/(2m)$ and $-b/m$.

9. Consider a function q of p, and let $P = \ln(p)$ so that $p = e^P$. Then define a new function Q of P by the formula $Q(P) = \ln(q(e^P))$. Assume that p and q take on positive values, and that the derivative q' exists. By means of the chain rule for differentiating composite functions, prove that

$$\frac{dQ}{dP} = \frac{p}{q(p)} \cdot q'(p).$$

10. Let f be a positive, differentiable function, defined on some interval I in $(0, \infty)$; furthermore, denote by $\eta_f(x)$ the elasticity of f at x, and choose a point x_0 in I. Prove that for every x in I,

$$f(x) = f(x_0) \cdot \exp\left(\int_{x_0}^{x} \frac{\eta_f(t)}{t}\, dt\right).$$

7.3 Third Level: Directed Applied Reading

Choose and read one of the following articles, 11–15, and answer the corresponding questions. Keep in mind these questions do not require a deep scrutiny of every mathematical detail in the references.

11. Study the importance of the demand for cigarettes in [Baltagi and Levin 1986].

    11.1 What numerical value did they obtain for the price elasticity of demand?
    11.2 Why did they need this value, and what conclusions did they reach?

12. Look at the analysis of export elasticities in [Jager and de Jong 1984].

    12.1 State their numerical value for the price elasticity of exports.
    12.2 What conclusion did they draw from their research?

13. Read Kleiman's comments on the legalization of marijuana in [Kleiman 1986], and Zeese's reply in [Zeese 1986].

    13.1 Kleiman fears that legalizing marijuana might triple consumption. Suppose that the demand equation $q = 15.4 \cdot p^{-1.013}$ holds [Nisbet and Vakil 1972]; by how much should the price decrease for consumption to triple?
    13.2 Assume instead that the alternate demand equation $q = -0.225p + 3.74$ holds, and assume a current price of \$10 per ounce [Nisbet and Vakil 1972]; can consumption triple?
    13.3 To avoid a huge increase in the consumption of marijuana Kleiman advocates raising taxes on wine and beer to reduce alcohol consumption and lower the rate of violent crime. Suppose that the formula $q = 1/p$ expresses the

demand for beer [Hogarty and Elzinga 1972]; by how much would the consumption of beer decrease after the imposition of an additional tax of, say, 20%?

14. Consult the paper [White and Luksetich 1983]. Is the demand for prostitutes' services elastic or inelastic? Justify your answer.

15. This reading provides an example of elasticity from river management in fishery and aquatic science; even though this use of elasticity does not refer to price elasticity of demand, mathematically the elasticity is the same as in economics. First, read the article about the sensitivity of salmonid populations in [Evans and Dempson 1986].

   15.1 Describe specific examples of elasticity from this reference; instead of price and quantity, what variables do they involve?
   15.2 Summarize the authors' main conclusions.
   15.3 In their research (pp. 864–865), the authors needed a product formula for elasticity. In words, the elasticity of a product of two functions equals the sum of their respective elasticities. Specifically, with $f \cdot g$ denoting the product of two functions f and g and with η_{fg} representing the elasticity of $f \cdot g$,

   $$\eta_{fg}(p) = \eta_f(p) + \eta_g(p).$$

   Prove this formula. (**Hint**: use the product rule for derivatives.)

8. Model Examination

Disclaimer: Since an examination tests students' ability to solve *new* problems, then the actual course examination may differ somewhat from this model.

Problem 1. State a definition of the "elasticity" of a function.

Problem 2. Jeffrey E. Harris estimated the price elasticity of demand for cigarettes about at -0.44 [Harris 1980]. Denote by p the price of cigarettes in dollars per pack, and by q the average smoker's consumption in packs per day. For instance, $p_1 = 0.52$ and $q_1 = 1$ are typical values for 1975.

   2.1 Determine the coefficients C and η such that the demand curve given by $q = C \cdot p^\eta$ passes through $(p_1, q_1) = (0.52, 1)$ with constant elasticity -0.44.

2.2 By approximately what percentage does the consumption of cigarettes change if the price increases by, say, 15%?

Problem 3. According to Huang, Siegfried, and Zardoshty, the price elasticity of demand for regular coffee remains approximately constant at -0.16, regardless of price [Huang et al. 1980].

3.1 By approximately what percentage should the *price* of coffee increase for the *consumption* to decrease by, say, 10%?
3.2 With a price increase, do sales revenues increase or decrease?

Problem 4. Let f be a positive, differentiable function, defined on an interval I on the positive real axis $(0, \infty)$. Prove that if the elasticity of f remains constant on I, then there exist two numbers K and c such that $f(x) = K \cdot x^c$ for every x in I.

Problem 5. Consider the straight line with equation $q(p) = mp + b$, with negative slope $m < 0$ and positive vertical intercept $b > 0$. For every p between 0 and $-b/m$ ($0 < p < -b/m$), let $A(p)$ denote the area of the rectangle with vertices at $(0,0)$, $(p,0)$, $(p, q(p))$, and $(0, q(p))$. Prove that $A(p)$ reaches its maximum at $p = -b/(2m)$.

9. Solutions to All the Exercises

9.1 First Level: Routine Exercises

1.1 $\ln(c) = 0.2870 \cdot \ln(p) + 8.117$.
1.2 $\eta(p) = 0.2870$ for every p.
1.3 $\Delta\%c \approx \eta(p) \times (\Delta\%p) = 0.2870 \times 0.05 = 0.01435$: about 1.4%.

2.1 $\eta(p) = 0.3494$ for every p.
2.2 $\Delta\%C \approx \eta(p) \times (\Delta\%p) = 0.3494 \times 0.05 = 0.01747$: about 1.7%.
2.3 The rate of property crimes, c, is larger: $c(57.51) = 3351 \times (57.51)^{0.2870} \approx 10{,}721$, whereas $C(57.51) = 207.8 \times (57.51)^{0.3494} \approx 856$.

3.1 $(P_1, Q_1) = (\ln(0.160), \ln(9.556)) = (-1.833, 2.257)$, and $(P_2, Q_2) = (\ln(0.236), \ln(5.055)) = (-1.444, 1.620)$.
3.2 $Q = -1.638P = 0.7454$.
3.3 $q = \exp(-1.638 \cdot \ln(p) - 0.7454) = 0.4745 \cdot p^{-1.638}$.
3.4 $\eta(p) = -1.638$ for every p.
3.5 $\Delta\%q \approx \eta(p) \times (\Delta\%p) = -1.638 \times 0.20 = -0.3277$: about 33%.

21

4.1 $\eta(p) = (p/q(p))q'(p) = (p/(-0.225p + 3.74))(-0.225) = -0.225p/(-0.225p + 3.74)$.

4.2 $\eta(10) = (-0.225 \times 10)/(-0.225 \times 10 + 3.74) = -1.510$.

4.3 Since $\Delta\%q \approx \eta(p) \times (\Delta\%p)$, then $\Delta\%p \approx (\Delta\%q)/\eta(p) = -0.05/-1.51 = 0.033$, or 3.3%.

4.4 Solve $(-0.255p)/(-0.225p + 3.74) = -1$, which gives $p^* \approx 8.311$.

5.1 From $\eta = d\ln(q)/d\ln(p)$ follows $M = \eta = -0.79$. Since the curve passes through $(4.85, 2)$, then $\ln(2) = -0.79 \cdot \ln(4.85) + B$, and hence $B = \ln(2) + 0.79 \cdot \ln(4.85) = 1.941$. Therefore $q = \exp(-0.79 \cdot \ln(p) + 1.941) = e^{1.941}p^{-0.79} = 6.96 \cdot p^{-0.79}$.

5.2 $\Delta\%p = (5.00 - 4.85)/4.85 = 0.0309$: about 3%. $\Delta\%q \approx \eta(p) \times (\Delta\%p) = -0.79 \times 0.0309 = -0.0244$: about 2.4%. Total expenditures increase because $|\eta| = 0.79 < 1$ (the demand is inelastic).

5.3 Suppose that the straight line with equation $q = mp + b$ passes through the point $(4.85, 2)$. Then $2 = m \times 4.85 + b$. Furthermore, if the elasticity equals -0.79 at the point $(4.85, 2)$, then

$$-0.79 = \eta(4.85) = \frac{4.85}{m \cdot 4.85 + b} \cdot m,$$

and multiplying both sides by the denominator gives $4.85m = -0.79(4.85m + b) = -3.815m - 0.79b$. Consequently, m and b satisfy two linear equations in two unknowns:

$$\begin{cases} 8.6815m + 0.79b = 0 \\ 4.85m + b = 2 \end{cases}$$

Therefore $m = -0.326$ and $b = 3.58$, hence $q = -0.326p + 3.58$.

9.2 Second Level: Mathematical Problems

6.

$$\frac{(g(w) - g(x))/g(x)}{(w - x)/x}$$

$$= \frac{((mw + b) - (mx + b))/(mx + b)}{(w - x)/x}$$

$$= \frac{m(w - x)/(mx + b)}{(w - x)/x} = \frac{mx}{mx + b} = \eta_g(x).$$

Thus the arc elasticity does not depend on w, but only on x; moreover, it always equals the elasticity, $\eta_g(x)$, regardless of w. Hence taking the limit as w tends to x does not change the result and gives $\eta_g(x)$. Finally, the examples in the text show that the function given by $f(x) = 1/x$ has elasticity -1, but the arc elasticity is -1.25 from $x = 3$ to $w = 2.4$.

7. Suppose that

$$\eta_g(x) = \frac{(g(w) - g(x))/g(x)}{(w - x)/x}$$

for every w and one fixed x. Then solving for $g(w)$ gives $g(w) - g(x) = \eta_g(x) \cdot (w - x) \cdot g(x)/x$ and

$$g(w) = (\eta_g(x) \cdot g(x)/x) \cdot w + g(x) \cdot (1 - \eta_g(x)),$$

the graph of which (in terms of $g(w)$ versus w) is a straight line with slope $m = (\eta_g(x) g(x)/x)$ and vertical intercept $b = g(x) (1 - \eta_g(x))$. In other words, since x remains fixed, then $g(w) = mw + b$.

8. Recall that $\eta_q(p) = (p/q(p))q'(p) = (p/(mp + b))m = mp/(mp + b)$.

  8.1 If $0 < p < -b/(2m)$, multiplying these inequalities by the negative number m gives $0 > mp > -b/2$. Next, adding b produces $mp + b > -b/2 + b = b/2 > 0$; therefore $mp + b > b/2 > 0$ and $mp > -b/2$. Taking the quotient of these two inequalities yields $0 > mp/(mp + b) > (-b/2)/(b/2) = -1$, which means that $0 > \eta_q(p) > -1$. (8.3) is similar.
  8.2 $\eta_q(-b/(2m)) = (m(-b/(2m)))/(m(-b/(2m)) + b) = (-b/2)/-b/2 + b) = (-b/2)/(b/2) = -1$.

9.

$$\frac{dQ}{dP} = \frac{d}{dP}\ln(q(e^P)) = \frac{1}{q(e^P)}q'(e^P)e^P = \frac{1}{q(p)}q'(p)p$$

$$= \eta_q(p).$$

10. Since $\eta_f(x) = xf'(x)/f(x)$, then $f'(x)/f(x) = \eta_f(x)/x$. Integrating from x_0 gives

$$\ln(f(x)) - \ln(f(x_0)) = \int_{x_0}^{x} \frac{f'(t)}{f(t)} dt = \int_{x_0}^{x} \frac{\eta_f(t)}{t} dt.$$

Thus, $\ln(f(x)) = \ln(f(x_0)) + \int_{x_0}^{x}(\eta_f(t)/t)\,dt$. Taking the exponential of both sides finally gives

$$f(x) = f(x_0) \cdot \exp\left(\int_{x_0}^{x} \frac{\eta_f(t)}{t}\,dt\right).$$

9.3 Third Level: Directed Applied Reading

11. Baltagi and Levin obtained the value -0.2 for the price elasticity of demand for cigarettes [Baltagi and Levin 1986]. They also measured such other parameters as elasticity with respect to income (about zero in this case) and elasticities within and between individual states. They concluded that a tax on cigarettes does not significantly lower consumption, but it provides an effective way to generate tax revenues, because the demand is inelastic. They also found that anti-smoking warning messages may slightly reduce consumption.

12. Jager and de Jong estimated the price elasticity of exports from the Netherlands about at -1.55 [Jager and de Jong 1984]. They pointed out that in spite of this apparent elasticity ($|\eta| = 1.55 > 1$), other macroeconomic effects make the volume of exports much less sensitive to price differences between countries.

13.1 Suppose that the consumption of marijuana triples, from q_1 to $q_2 = 3q_1$. Solving for p the demand equation $q = 15.4 \cdot p^{-1.013}$ gives $p = 14.8 \cdot q^{-0.987}$, hence the price p must decrease by a factor of 3:

$$\begin{aligned}p_2 &= 14.8 \cdot q_2^{-0.987} = 14.8(3q_1)^{-0.987}\\ &= 3^{-0.987}14.8 \cdot q_1^{-0.987}\\ &= \tfrac{1}{3}p_1.\end{aligned}$$

The approximation $\eta_q(p) \approx (\Delta\%q)/(\Delta\%p)$ is no longer accurate enough for $\Delta\%q$ as large as 300%.

13.2 With the affine demand curve, $q = -0.225p + 3.74$, consumption reaches a maximum with $q_0 = 3.74$ at $p_0 = 0$, while the current consumption amounts to $q_1 = 1.49$ at $p_1 = 10$. Therefore, under this hypothesis, consumption could not triple.

13.3 Since the demand for beer is unit elastic ($\eta = -1$), a price increase of 20% would reduce the consumption by approximately $\Delta\%q \approx (-1) \times (\Delta\%p) = -20\%$. More exactly, if $p_2 = 1.20p_1$, then $q_2 = 1/p_2 = 1/(1.20p_1) = 0.83q_1$: consumption decreases by 17%.

14. The demand for prostitutes' services is inelastic. Indeed, harassing the customers reduces the quantity demanded, which tends to decrease the price; a price drop with an inelastic demand reduces the total revenues (recall that a price increase with an inelastic demand increases the revenues). Harassing the suppliers instead of the customers would reduce the quantity supplied and hence would raise both the price and the revenues, or total amount spent on prostitutes' services. In this sense, harassing the buyers appears more "effective" than harassing the prostitutes. If the demand were elastic, then the opposite would happen.

15.1 The research by Evans and Dempson concerns the abundance of salmon in a river, and it involves such variables as the fractions, or probabilities, of survival of eggs, grilse, parr, and smolts (instead of price or income), and the abundance of fish available (instead of the quantity sold). The authors' paper estimates the elasticities of the *number of fish at a specific age* with respect to the *probabilities of survival at previous stages*.

15.2 Having determined that the population of salmon is more sensitive to certain factors than others (by comparing various elasticities), Evans and Dempson conclude that measures to protect the habitat seem more reliable than the regulations of the fishery industry.

15.3

$$\eta_{fg}(x) = \frac{x}{f(x)g(x)} \cdot \frac{d}{dx}(f(x)g(x))$$

$$= \frac{x}{f(x)g(x)} \cdot (f'(x)g(x) + f(x)g'(x))$$

$$= \frac{xf'(x)g(x)}{f(x)g(x)} + \frac{xf(x)g'(x)}{f(x)g(x)}$$

$$= \frac{x}{f(x)}f'(x) + \frac{x}{g(x)}g'(x) = \eta_f(x) + \eta_g(x).$$

10. Solutions to the Model Examination

Problem 1. Let f be a positive, differentiable function, defined on an interval I of the positive real axis $(0, \infty)$. Then the elasticity of f is a function $\eta_f \colon I \to \mathbf{R}$ defined by either of the following formulae.

(See **Definition 3**, for instance.)

$$\eta_f(x) = \lim_{w \to x} \frac{(f(w) - f(x))/f(x)}{(w-x)/x}; \quad \eta_f(x) = \frac{x}{f(x)} f'(x);$$

$$\eta_f(x) = \frac{d \ln(f(x))}{d \ln(x)}.$$

(Only the formula in terms of logarithms requires the assumption that both x and $f(x)$ be positive.)

Problem 2.
2.1 $\ln(q) = -0.44 \cdot \ln(p) + b$ and $\ln(1) = -0.44 \cdot \ln(0.52) + b$. Hence $b = 0.44 \cdot \ln(0.52) = -0.2878$ and finally $q = 0.75 \cdot p^{-0.44}$, as in **exercise 5.1**.
2.2 $\Delta\%q \approx \eta_q(p) \times (\Delta\%p) = (-0.44) \times (0.15) = 0.066$: about 6.6%, as in **exercises 1.3, 2.2**, and **3.5**.

Problem 3.
3.1 $\Delta\%p \approx (\Delta\%q)/\eta_q(p) = -0.10/-0.16 = 0.62$: about 62%, as in **exercise 4.3**.
3.2 The revenues increase because the demand for coffee is inelastic.

Problem 4. Proceed as in **exercise 10**. Since $xf'(x)/f(x) = \eta_f(x) = c$, then $f'(x)/f(x) = c/x$, and integrating each side gives $\ln(f(x)) = c \cdot \ln(x) + k$. Then put $K = e^k$ so that $k = \ln(K)$, which gives $f(x) = K \cdot x^c$.

Problem 5. The area $A(p) = pq$ reaches its maximum where $\eta_q(p) = -1$, which occurs at the midpoint $p = -b/(2m)$, as proved in the text and in **exercise 8**. Alternatively, the formula $A(p) = pq = p \cdot (mp + b) = mp^2 + bp$ shows directly that $A(p)$ attains its maximum at $p = -b/(2m)$.

Other methods of solution exist, and other approximations may be satisfactory.

11. References

Baltagi, Badi H., and Dan Levin. 1986. Estimating dynamic demand for cigarettes using panel data: the effect of bootlegging, taxation, and advertising reconsidered. *The Review of Economics and Statistics* 68(1):148–155.

Brown, George F., Jr., and Lester R. Silverman. 1973. *The Retail Price of Heroin: Estimation and Applications*. Washington, DC: The Drug Abuse Council, Inc.

Evans, Geoffrey T., and J. Brian Dempson. 1986. Calculating the sensitivity of a salmonid population model. *Canadian Journal of Fishery and Aquatic Science* 43(4):863–868.

Harris, Jeffrey E. 1980. Taxing tar and nicotine. *The American Economic Review* 70(3):300–311.

Hogarty, T. F., and K. G. Elzinga. 1972. The demand for beer. *The Review of Economics and Statistics* 54(2):195–198.

Huang, Cliff J., John J. Siegfried, and Farangis Zardoshty. 1980. The demand for coffee in the United States, 1963–1977. *Quarterly Review of Economics and Business* 20(2):36–50.

Jager, Hank, and Eelke de Jong. 1984. On export performance and export elasticities: a macroeconomic approach. *De Economist* 32(2):224–231.

Kleiman, Mark. 1986. Marijuana prohibition doesn't pass Gramm-Rudman test. *The Wall Street Journal* 114 (May 8, 1986):28.

Kohli, Ulrich. 1986. Robert Giffen and the Irish potato: note. *The American Economic Review* 76(3):539–542.

Nisbet, C.T., and F. Vakil. 1972. Some estimates of price and expenditure elasticities of demand for marijuana among UCLA students. *The Review of Economics and Statistics* 54(4):473–475.

Silverman, Lester P., and Nancy L. Spruill. 1977. Urban crime and the price of heroin. *Journal of Urban Economics* 4(1):80–103.

Simon, Julian L. 1966. The price elasticity of liquor in the U.S. and a simple method of determination. *Econometrica* 34(1):193–205.

Suits, Daniel B. 1979. The elasticity of demand for gambling. *Quarterly Journal of Economics* 93(1):155–162.

Wales, Terrence J. 1968. Distilled spirits and interstate consumption effects. *The American Economic Review* 57(4):853–863.

White, Michael E., and William A. Luksetich. 1983. Heroin: price elasticity and enforcement strategies. *Economic Inquiry* 21(4):557–564.

Zeese, Kevin. 1986. What if pot were legal? *The Wall Street Journal* 114 (May 20, 1986):29.

UMAP

Modules in Undergraduate Mathematics and its Applications

Published in cooperation with the Society for Industrial and Applied Mathematics, the Mathematical Association of America, the National Council of Teachers of Mathematics, the American Mathematical Association of Two-Year Colleges, The Institute of Management Sciences, and the American Statistical Association.

Module 675

The Lotka-Volterra Predator-Prey Model

James Morrow

INTERMODULAR DESCRIPTION SHEET:	UMAP Unit 675
TITLE:	THE LOTKA-VOLTERRA PREDATOR-PREY MODEL
AUTHOR:	James Morrow SummerMath Mount Holyoke College 302 Shattuck Hall South Hadley, MA 01075
MATHEMATICAL FIELD:	Precalculus
APPLICATION FIELD:	Biology
TARGET AUDIENCE:	Students in a precalculus course
ABSTRACT:	This module describes and analyzes qualitatively a simplified version of the predator-prey model attributed to Lotka and Volterra. Deductions are made concerning the size of populations based on information about their percentage growth rates. The module describes a nonstandard and stimulating way of illustrating the power and utility of combining geometry and algebra.
PREREQUISITES:	1. to solve a linear inequality 2. to graph a linear equation
STUDENT OBJECTIVES:	1. to be able to sketch a plausible two-species population trajectory based on an algebraic description (of the Lotka-Volterra type described in the module) of the species' percentage growth rates; 2. to be able to sketch a population trajectory based on the population vs. time graphs for each of the two species; 3. to be able to express verbally the information conveyed by a two-species population trajectory and the value and limitations of such trajectories.

© Copyright 1987 by COMAP, Inc. All rights reserved.

COMAP, 60 Lowell Street, Arlington, MA 02174 (617) 641-2600

The Lotka-Volterra Predator-Prey Model

James Morrow
SummerMath
Mount Holyoke College
302 Shattuck Hall
South Hadley, MA 01075

Table of Contents

1. THE LOTKA-VOLTERRA MODEL 1
   1.1 Introduction 1
   1.2 Examples and Notation 1
   1.3 Description of Growth Rates 2
   1.4 Assumptions of the Model 3
   1.5 Conclusions of the Model 4
   1.6 Concluding Remarks 9
2. EXERCISES 10
3. SOLUTIONS TO SELECTED EXERCISES 15
4. REFERENCES 18

Modules and Monographs in Undergraduate Mathematics and Its Applications (UMAP) Project

The goal of UMAP was to develop, through a community of users and developers, a system of instructional modules in undergraduate mathematics and its applications to be used to supplement existing courses and from which complete courses may eventually be built.

The Project was guided by a National Advisory Board of mathematicians, scientists, and educators. UMAP was funded by a grant from the National Science Foundation and is now supported by the Consortium for Mathematics and Its Applications (COMAP), Inc., a nonprofit corporation engaged in research and development in mathematics education.

COMAP STAFF

Solomon A. Garfunkel	Executive Director, COMAP
Laurie W. Aragon	Business Development Manager
Philip A. McGaw	Production Manager
Nancy Hawley	Copy Editor
Annemarie S. Morgan	Administrative Assistant

1. The Lotka-Volterra Model

1.1 Introduction

In the 1920s A.J. Lotka developed a mathematical model for the interaction between two species [Lotka 1925]. This model was worked out independently and in more detail a short time later by the mathematician Vito Volterra. Lotka and Volterra wished to understand the population dynamics involved in a simple system that involves only a single predator species and a single prey species. By ignoring such things as the variability of individuals of each species, variation in the environment over time, and the effects of other species, they hoped to discover some of the essential properties of the interaction between the two species and to understand the mechanisms involved in the cyclical behavior of their population levels. While it is still an open question as to whether they achieved their goals, the mathematical approach that they took stimulated many investigators and has changed the way that ecological systems are studied.

"...the mathematical approach that they took... has changed the way that ecological systems are studied."

In this module we shall examine the assumptions and conclusions of a slightly modified version of the Lotka-Volterra model. A subsequent related module contains a model that describes the dynamics of competing species.

Before reading on, take a few minutes to jot down some ideas and questions you have about a predator-prey model. For example, you might make a preliminary decision about what variables should be included; you might decide on what you hope the model would accomplish; and you might ask why one would try to form such a model.

1.2 Examples and Notation

"The variety of examples to be described by a single mathematical model indicates how ambitious that model is."

Some examples of predator-prey systems that might be described by the Lotka-Volterra model include the hawk-sparrow, lynx-hare, wolf-caribou, pitcher plant-fly, and squirrel-acorn systems. The variety of examples to be described by a single mathematical model indicates how ambitious that model is. To make the ideas a little more concrete, we shall assume that the prey is a population of hare and the predators are Canadian lynx. A wealth of data has been collected, primarily from the pelt counts of human predation, which could provide a way of validating the Lotka-Volterra model [Elton and Nicholson 1942].

Out of the complexity of the population dynamics of the real predator-prey system, the model abstracts in an explicit way only four variable quantities: the number of predators, the number of prey, and the percentage growth rates of each. We use the following notation.

Prey (hare): population size = H

percentage growth rage = r_H

Predator (lynx): population size = L

percentage growth rage = r_L

There is another variable quantity that enters the model implicitly: time. The model assumes that time is independent of the population sizes and rates and that it "flows along" in a continuous way. The preceding four explicitly labeled quantities represent variables that are assumed to change over time, and hence describe how the population sizes change with time. It is assumed that each can be specified at each point in time and thus is dependent (although in an implicit way) on time. Section 1.5 contains a bit more discussion of this point.

How does this choice of quantities compare with your ideas about a predator-prey system? Many different approaches to a model are possible, so don't be discouraged if your ideas seem radically different from those presented here.

1.3 Description of Growth Rates

The growth of populations can be described in several ways. For example, one reads that the world population in the mid-1960s was growing at an absolute rate of 180,000 people per day, or that world population grew by 1.7% per year from 1950 to 1960. The latter way of describing a rate of growth differs from the former not only in the unit of time used (day in the former case and year in the latter), but more fundamentally, in the use of a percentage rate rather than an absolute rate.

Question: If a certain population grows at a rate of 1.7% per year for a year, how many additional mouths are there to feed at the end of the year?

The idea that a population is growing at 1.7% per year cannot alone determine the additional number of people each year. To determine that additional number requires the use of a base population figure. Thus a population of 1,000 will grow to 1,017 in a year's time, while a population of 2,000 will grow to 2,034—each at a rate of 1.7% per annum. Percentage rates of growth are used for populations because they have generally been more stable figures than absolute rates of growth. A constant, positive percentage rate describes a changing (increasing) absolute rate of growth. Even percentage rates change—it is estimated that up to 1750, world population grew at 0.1% per year, from 1750 to 1900 at 0.5% per year, from 1900 to 1950 at 1.0% per year, and from 1950 to 1960 at 1.7% per year [Young 1968]. The corresponding absolute rates changed even more dramatically.

1.4 Assumptions of the Model

We first consider the situation of the hare without any lynx to prey on them. We assume that the hare population, in the absence of lynx, will grow with a constant, positive percentage rate. Symbolically, $r_H = a$, where a is a positive constant (this is termed the "intrinsic" percentage rate—it applies only in the absence of lynx). More specifically, $r_H = 0.05$ means that the intrinsic growth rate is 5% per year. Rather than choosing a *specific* number to represent the growth rate, the constant a is used so that the model can be more widely applied, and also so that general conclusions may be drawn which are independent of the particular value that a may have.

We also assume that the lynx will have the effect of decreasing the intrinsic percentage growth rate in direct proportion to the lynx population size. Combining the intrinsic growth rate with the effect of the predation of the lynx population, we assume that

$$r_H = a - b * L, \tag{1}$$

where a and b are positive constants. (The symbol " $*$ " is used to denote multiplication.) The symbol b represents the constant of proportionality involved in the lynx's effect on the growth rate of the hare population. More precisely, b is the percentage kill rate of hare per lynx, and $b * L$ is the percentage kill rate of hare. It should be noted that the more effective the lynx are in killing the hare, the larger the value of b is. If we knew more about the rate at which lynx killed hare, we would be able to replace b by some specific number. On the other hand, the model has more generality by using the unknown but fixed constant b.

In a parallel way, we assume that the lynx will die out in the absence of hare in such a way that their percentage growth rate is a negative constant: specifically, $r_L = -c$, where c is a positive constant (the "intrinsic" percentage death rate—it applies only in the absence of the lynx' food source, the hare).

The presence of hare, however, will increase the intrinsic percentage growth rate, it is assumed, in direct proportion to their population size, so that

$$r_L = -c + d * H, \qquad (2)$$

where c and d are positive constants. Just as b represents how effectively the lynx kill the hare, d represents how effective the lynx' predation is in increasing the lynx population size.

Equations (1) and (2) describe our assumptions about the lynx-hare system in a precise mathematical way. The symbols L, H, r_L, and r_H all represent variables which change over time, while the symbols a, b, c, and d represent constants, numbers that are fixed but unknown. These constants are called the parameters of the model. Just as the *variables* describe the population sizes and rates of growth, which vary with time, the *parameters* describe certain characteristics of the populations, which vary from one predator-prey system to another. Thus, for example, the greater the intrinsic percentage growth rate of a prey population, the greater the value of the parameter a. How does the intrinsic death rate of lynx vary with the values of the parameter c?

> "...the parameters describe certain characteristics of the populations, which vary from one predator-prey system to another."

1.5 Conclusions of the Model

We have tried to make a case for the plausibility of the assumptions made in the previous section. Perhaps you have some difficulties with them. Well, you're not alone! Nearly everyone would agree that those simple assumptions capture only a small part of a complex world. As a measure of how far off those assumptions might be, we shall now examine some mathematical conclusions that can be drawn from them.

What are these conclusions? Keep in mind that we are not asking the question of what happens to actual populations of hare and lynx, but, rather, about the theoretical population sizes symbolized by H and L. What are the long-term tendencies for the "hare" and "lynx" populations? To answer these questions it is critical to know when the percentage growth rates are positive and when they are negative, for this will determine whether the populations are increasing or decreasing.

> "What are the long-term tendencies for the 'hare' and 'lynx' populations?"

Rather than considering the model in complete generality, let us first look at a specific case. This will help us to see the general pattern. Suppose that the predator-prey system is governed by the equations: $r_H = 0.05 - 0.001 * L$; $r_L = -0.03 + 0.0002 * H$.

In addition, let us suppose that at a certain point in time there are 40 lynx and 250 hare. Then $r_H = 0.05 - 0.001 * 40 = 0.01$; and $r_L = -0.03 + 0.0002 * 250 = 0.02$.

What is most important to note is that both r_H and r_L are positive at these population levels. Also note that the percentage growth rate for lynx is double that for hare. We will try to guess what happens to each population over time. Since they both have positive rates of growth, both populations tend to increase in size. Since both populations are increasing in size, let us suppose, hypothetically, that they happen to grow to a point where there are 45 lynx and 265 hare. Then $r_H = 0.05 - 0.001*45 = 0.005$; and $r_L = -0.03 + 0.0002*265 = 0.023$.

We see that at these population levels the growth rates are still positive for each population, with the hare population growth slowing and the lynx increasing even faster. Let us try to trace what may happen a little further. Suppose, again hypothetically, that the lynx population reaches size 50 and the hare population reaches a size of 275. Then $r_H = 0.05 - 0.001*50 = 0.00$; and $r_L = -0.03 + 0.0002*275 = 0.025$.

At these population levels the percentage growth rate of the hare population reaches zero, while the lynx population still has a positive growth rate. It is the *lynx* population reaching a size of 50 that forces the *hare* growth rate to zero. Since the lynx population still has a positive growth rate, let us suppose that there are 51 lynx (and 275 hare). Checking the basic equations (1) and (2) again, we see that $r_H = -0.001$ and $r_L = 0.025$. Thus any time the lynx population exceeds 50, the hare population is forced to decline in number.

Rather than continue to play this "numbers" game, consider the whole process from a geometric point of view. There are three variables: the number of hare H, the number of lynx L, and time. We shall use a two-dimensional rectangular coordinate system, with L plotted on the horizontal axis and H plotted on the vertical axis. The variation over time will be shown by movement in this coordinate plane. What is critical to determine is where each of the two populations reaches a zero growth rate. We have already seen that the hare population reaches zero growth when the lynx number 50. When will the lynx population have a zero growth rate?

By solving the equation $-0.03 + 0.0002*H = 0$ for H, it can be seen that when there are 150 hare, the lynx growth rate is zero. These two pieces of information are noted by drawing two dashed lines: the vertical one to represent $L = 50$ and the horizontal one to represent $H = 150$. The coordinate plane is divided into four regions:

In **Figure 1** the coordinate system appears, along with the critical lines; while in **Figure 2** the scenario that was just described in words is depicted geometrically. Each single point plotted represents a possible combination of lynx-hare population levels, and the arrow attached to each point indicates the growth tendency at those population levels. (Up is positive growth for hare and right is

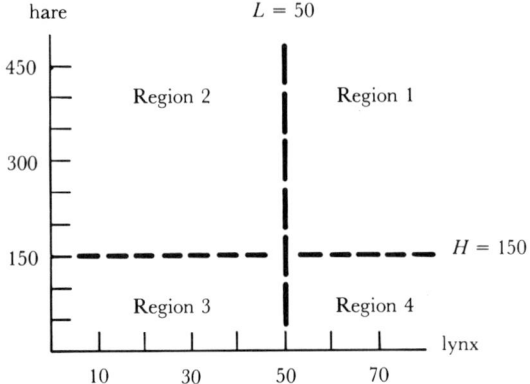

Figure 1.

positive growth for lynx.) Each of the four regions of the coordinate plane can be checked numerically to see what growth tendency is present. For example, in Region 4, where $L > 50$ and $H < 150$, we have $r_H < 0.00$ and $r_L < 0.00$; hence in Region 4 both populations are on the decline. Some typical growth tendencies are depicted in **Figure 3**.

Imagine now what may happen over time. A population pair starting at the point A pictured in **Figure 3** will tend to move upward and to the right. As it moves in this direction, the hare *rate* of growth tends to decrease (though it is still positive), and the lynx *rate* of growth tends to increase, which makes it seem likely that the population pair will eventually hit the line $L = 50$. If the pair reaches the vertical line $L = 50$, it then begins to move downward and to the right. During this period of time the hare population increased until the lynx population hit 50 and then decreased, while the lynx population was steadily increasing. The population pair continues to move downward and to the right until the hare popula-

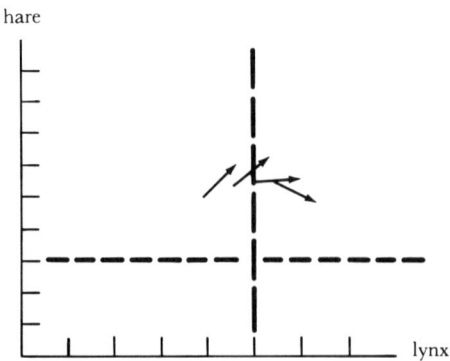

Figure 2.

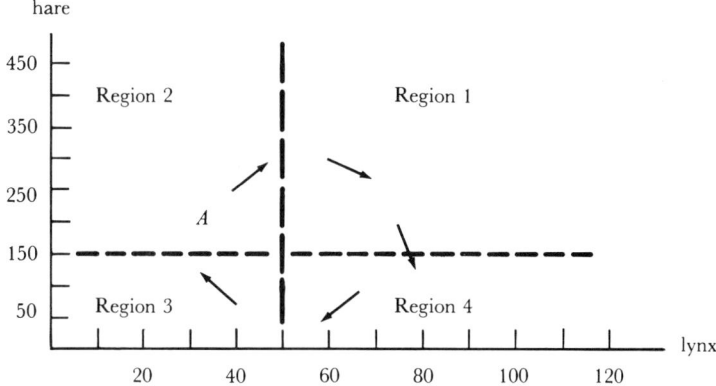

Figure 3.

tion reaches a size of 150, when it begins to move back to the left (corresponding to the now decreasing lynx population) while continuing down.

How long will the two populations continue to decrease in number? Although this question can't be answered in terms of time, it can be answered in the following way: the downward and leftward movement will continue until the lynx population size reaches 50 again (or until the hare population reaches zero—if the hare population ever reaches zero, the lynx population will also die out, since for $H = 0$, $r_L = -0.03$), at which time the motion changes to one of upward (increasing numbers of hare) and to the left. It is easy to guess what happens beyond this point; such a growth tendency persists until the hare population size reaches 150 once again (or until the lynx population reaches zero—see **Exercise 8**), at which point the situation is similar to that at point A, and the "cycle" is repeated. This wordy description of population dynamics is depicted very simply in **Figure 4**.

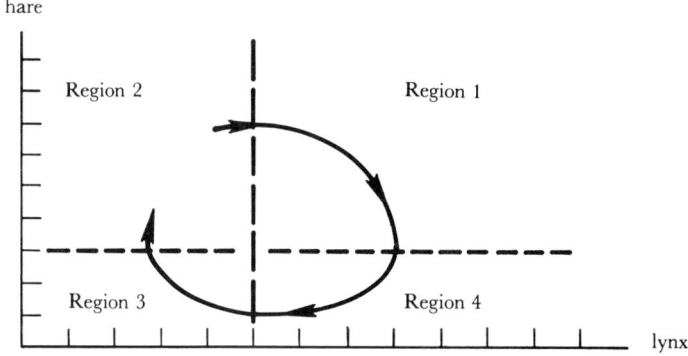

Figure 4.

The curve drawn in **Figure 4** illustrates the behavior resulting from the self-regulatory nature of the predator-prey relation modeled by the Lotka-Volterra equations. In "good" times for both species, both their numbers are increasing (Region 2). Eventually there are so many lynx that the hare population begins to decline (Region 1). For a while there are still enough hare left for the lynx population to continue to grow, but eventually there are few enough hare so that the lynx population begins to decline as well (Region 4). In time the lynx population becomes small enough so that once again the hare population begins to increase (Region 3), and eventually this increase in hare is enough to enable the lynx population to increase again (Region 2 again).

Let us return to the general situation. We have:

$$r_H = a - b * L \tag{1'}$$

and

$$r_L = -c + d * H, \tag{2'}$$

rather than the very specific

$$r_H = 0.05 - 0.001 * L \tag{1}$$

and

$$r_L = -0.03 + 0.0002 * H. \tag{2}$$

Thus where before the hare population growth rate was positive for $L < 50$ (and negative for $L > 50$), now r_H is positive for $L < a/b$. This new, and more general, inequality is obtained by solving the inequality $a - b * L > 0$ for L. In the special case, the lynx population growth rate r_L was positive for $H > 150$, while in the general case r_L is positive when H is greater than the quantity c/d. (Solve $-c + d * H > 0$ for H.) We proceed as before with a rectangular coordinate system, this time drawing the dashed vertical line $L = a/b$ and the dashed horizontal line $H = c/d$. (Of course we don't actually assign specific values to the parameters $a, b, c,$ and d, so our placement of these critical lines is rather arbitrary. No matter—the value of the pictures is of a *qualitative* rather than quantitative nature.) See **Figure 5**.

The geometric analysis of the general case is also easy, now that the four critical regions of the plane have been identified. The population pair moves upward and to the right in Region 2, downward and to the right in Region 1, downward and to the left in Region 4, and upward and to the left in Region 3. The only thing different about the general case is the precise position of the critical

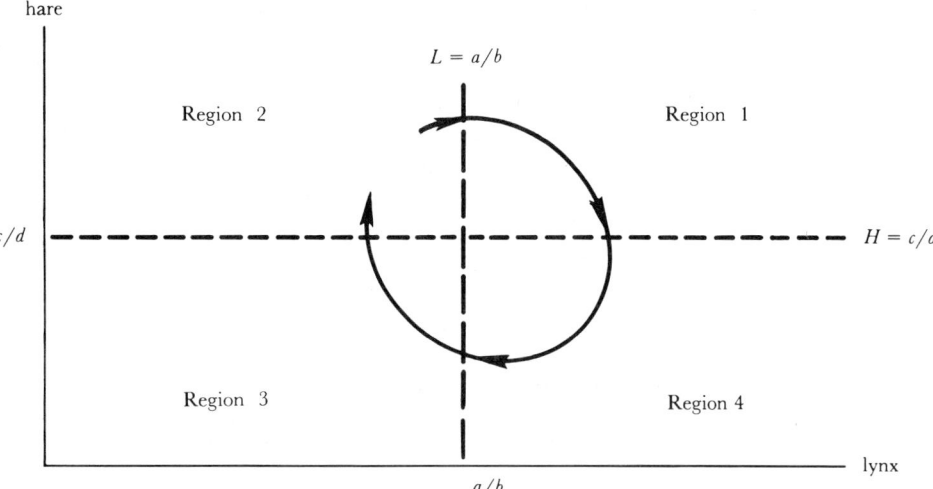

Figure 5.

dashed lines that divide up the plane. The qualitative nature of both is the same. The curve traced out over time by the population pair is called the population *trajectory*.

One of the subtle points about the model described here concerns the use of the term "percentage growth rate." Volterra and Lotka meant by this term an instantaneous rate, one that changes from instant to instant, rather than a rate that is obtained by averaging the changes in population levels over time. Such an instantaneous rate is modeled mathematically by the concept of derivative of a function, which may be studied in a calculus course. Using the "instantaneous" meaning of percentage growth rate and methods of differential calculus, Lotka and Volterra showed that the kind of smooth trajectory described above actually does result from the more sophisticated assumptions of their model. What is more striking is that they also showed that the trajectory is a *closed* curve in the plane. What this means is that the curve doesn't spiral around, but comes back around to itself, and then repeats. Thus the model results in a cyclical kind of behavior for each species which closely matches some natural predator-prey systems.

1.6 Concluding Remarks

In mathematical terms, the (modified) Lotka-Volterra model presented in this module illustrates the fruitful interaction of algebraic and geometric methods. The form of the model's assumptions is algebraic; and the description of zero, positive, and negative percentage growth is also algebraic, being in the form of an algebraic equation or inequality. When these equations and inequalities are considered geometrically, and the pair of populations is looked at as a point in the plane, a powerful tool develops—for then we "see" how the populations must vary over time if they are to satisfy the assumptions of the model. Such variation over time is seen in the form of a curve being traced out in a coordinate plane.

In ecological terms, the model provides a beginning for a quantitative analysis of natural systems. Such a simple model might best be considered as a basis for deeper understanding. Its value lies in whatever questions, ideas, and experiments it may stimulate and in the simple picture that we may carry around in our mind to remind us of our ideas about a system of interacting species. The interested reader can find many variations on the theme of the Lotka-Volterra model in [May 1976; Pielou 1969; and Wilson and Bossert 1971].

2. Exercises

1. What features are omitted from the Lotka-Volterra model that you think might be important? What quantities do you think might be included to make the model closer to the reality of a natural system involving a predator species and its prey species? Why do you think Lotka and Volterra failed to include such quantities?

2. Consider the predator-prey system $r_H = 0.08 - 0.002 * L$; $r_L = -0.04 + 0.0002 * H$.
   a. Determine the number of lynx for which the percentage growth rate of hare is zero.
   b. Determine the number of hare for which the percentage growth rate of lynx is zero.
   c. On the graph below, draw the critical lines that divide the lynx-hare plane into critical regions such as those described in the module.

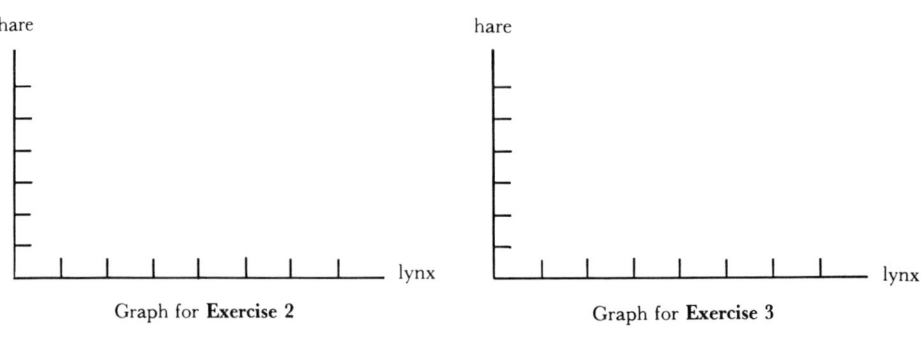

Graph for **Exercise 2**   Graph for **Exercise 3**

d. Plot the point in the plane corresponding to 35 lynx and 150 hare. Is the hare population growing or declining at that point? Is the lynx population growing or declining at that point?

3. Repeat **Exercise 2** for the system $r_H = 0.05 - 0.002 * L$; $r_L = -0.03 + 0.0001 * H$.

4. The two graphs below plot a prey species H against time and a predator species L against time. Draw a rough graph of the corresponding population trajectory of prey against predator.

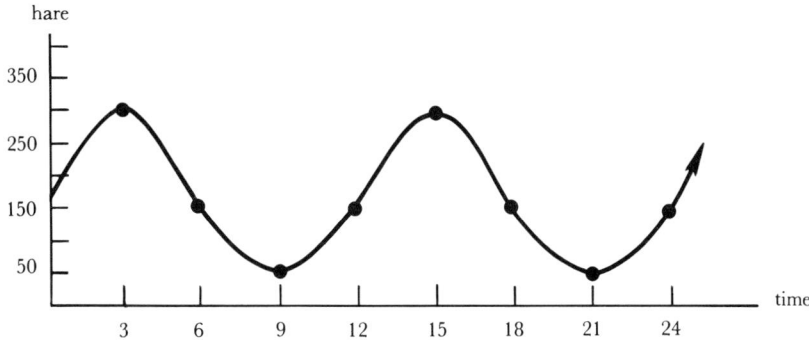

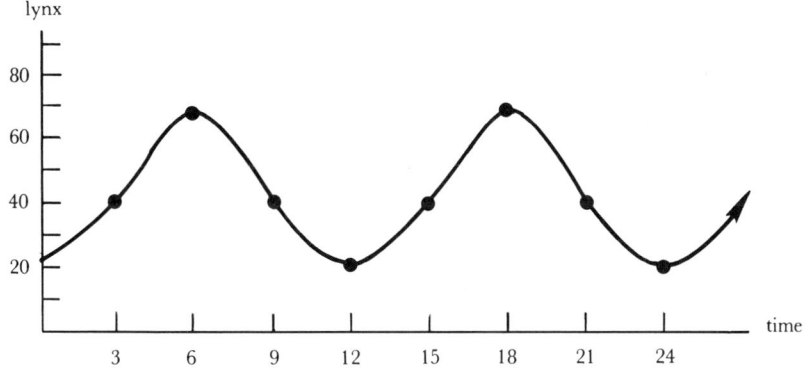

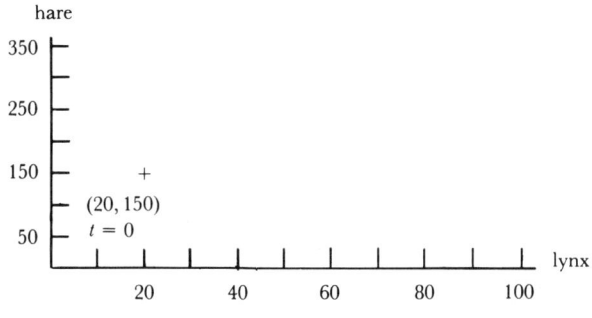

5. Repeat **Exercise 4** for the two time graphs drawn below.

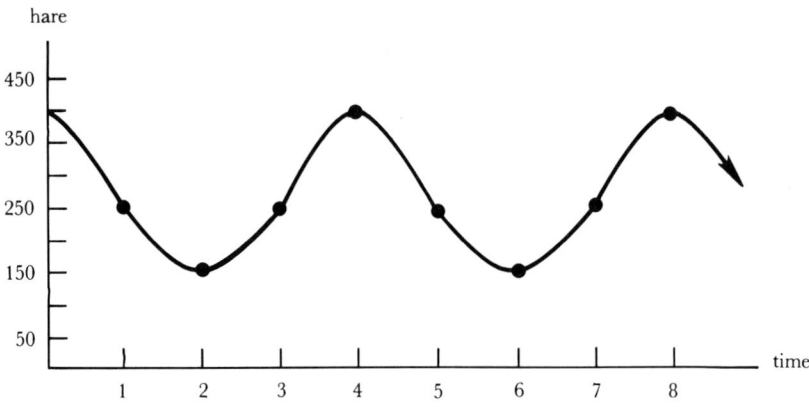

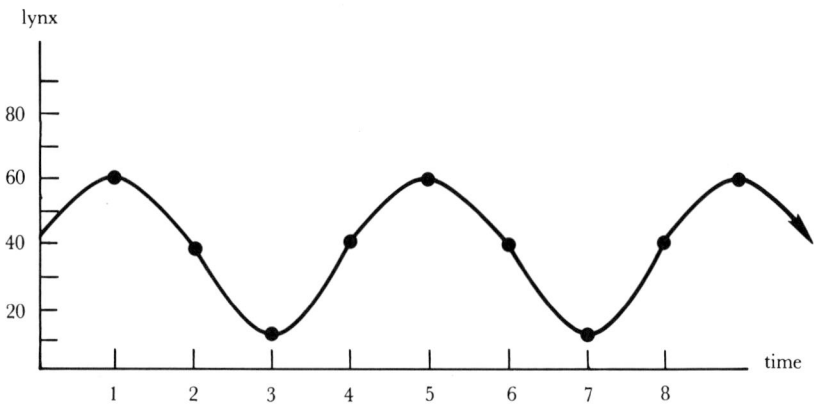

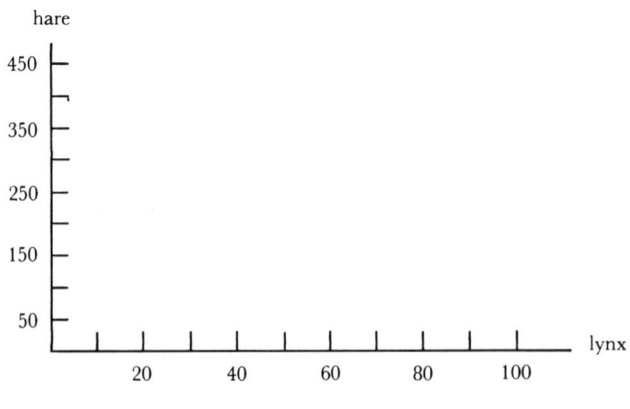

6. The figure below gives a possible population trajectory corresponding to the system of **Exercise 2**.

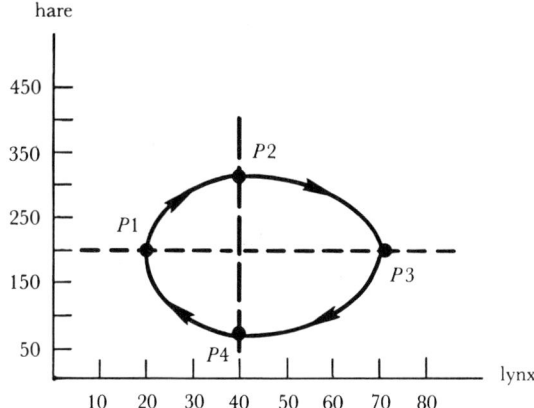

Draw a rough graph of the number of lynx against time. (Assume that the population pair is at point $P1$ at time $t = 0$ and that it takes the same amount of time for the pair to travel from $P1$ to $P2$ as from $P2$ to $P3$, as from $P3$ to $P4$, as from $P4$ to $P1$.)

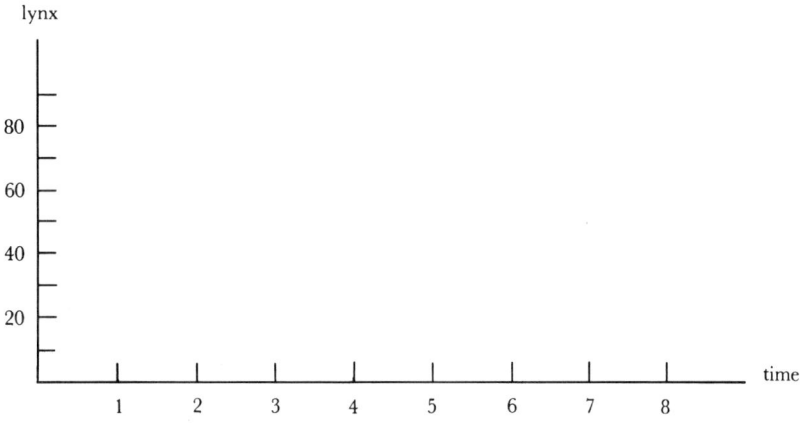

7. Repeat **Exercise 6** for the population trajectory that follows.

8. What happens to the hare population if the lynx population size ever reaches zero?

9. Put yourself in the place of a wildlife manager. Try to determine whether it is possible, according to the Lotka-Volterra model, to

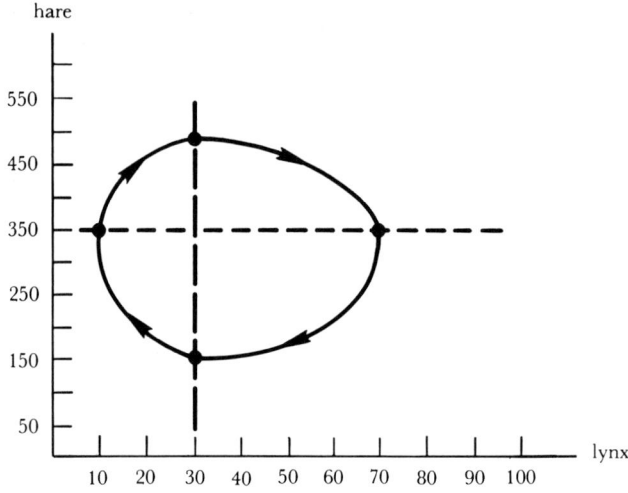

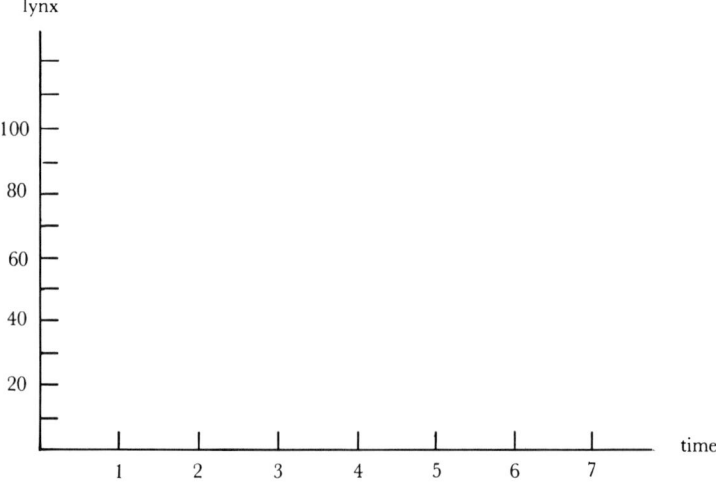

attain each of the management objectives listed below, considered separately, by removing lynx:
a. to increase the maximum size of the hare population.
b. to increase the minimum size of the hare population.
c. to decrease the maximum size of the lynx population.

10. Describe what it might mean for the value of the parameter b to increase. First consider the effect mathematically, then what the mathematical effect might stem from in terms of hare and lynx population characteristics.

11. Suppose there is a predator-prey system in which each of the populations grows by fairly distinct generations, most of the population change occurring in a short span of time. Then we might want to restrict our model to just the discrete points in time: an initial point in time, the time when the first generation is born, the time when the second generation is born, and so on. To make things as simple as possible we could assume that both populations have their new generations born at the same time, so that we could let $H(0)$ and $L(0)$ represent the initial populations; $H(1)$ and $L(1)$ represent the population sizes when the first generation is born; and, more generally, $H(n)$ and $L(n)$ represent the population sizes when the nth generation is born.
   a. Use the above notation to express the *actual increase* in the two population sizes as they go from their *initial population size* to the population size when the *first generation* is born.
   b. Use the above notation to express the *actual increase* in the two population sizes as they go from the time when the *n*th *generation* is born to the time when the $(n + 1)$th *generation* is born.
   c. Use the above notation to express the *percentage increase* in the two population sizes as they go from their *initial population* size to the population size when the *first generation* is born.
   d. Use the above notation to express the *percentage increase* in the two population sizes as they go from the time when the *n*th *generation* is born to the time when the $(n + 1)$th *generation* is born.
   e. Propose equations (a model) that describe how percentage increases in each population might be related to the actual population sizes.
   f. Write a computer program to test the equations you wrote in part e. above. You will need to use specific values for any parameters you introduced into the equation in part e.

3. Solutions to Selected Exercises

1. In addition to the variability of individuals, variation over time, and the effect of other species mentioned in the first paragraph of the module, one might try to take into account the distribution of species in space, variation in the behavior of individuals, food available (other than hare!), and the different age groups within a population. Without trying to change the essential characteristics of the model too much, I might try to include the variables time and amount of food available to the hare as a specific function of time. What is unfortunate is that the more realistic the number of factors considered and the more realistic the hypotheses are, the more difficult it is to draw any conclusions at all mathematically. This is the sort of trade-off that Lotka and Volterra faced (and that anyone trying to make a mathematical model faces). The mathematical difficulty introduced by the inclu-

sion of additional variables may account for Lotka and Volterra's failure to include such quantities as those listed here.

2. **a.** Set the growth rate for hare, r_H, equal to zero; i.e., consider $0.08 - 0.002 * L = 0$. Solving for L yields $L = 40$. Thus the hare growth rate is zero when there are 40 lynx.
   b. Similarly to part **a.**, one sets $-0.04 + 0.0002 * H = 0$, and solves for H to get $H = 200$. When there are 200 hare, the growth rate of the lynx population is zero.
   c.

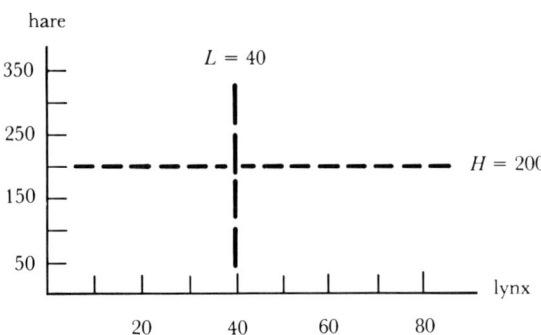

   d. For $L = 35$, $r_H = 0.08 - 0.002 * 35 = 0.01$, so the number of hare is increasing. For $H = 150$, lynx growth $= -0.04 + 0.0002 * 150 = -0.01$, so the number of lynx is decreasing.

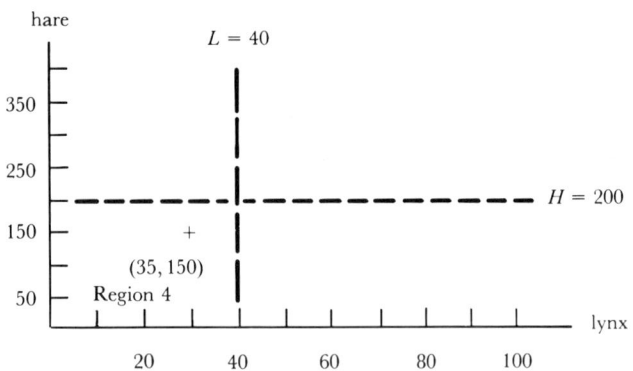

4. A beginning to the solution is given in the statement of the exercise: the point (20,150), which describes the population pair when $t = 0$, has been plotted. Reading the lynx and hare populations from the two graphs at times $t = 3, 6, 9$, and 12, we get the points (60,250), (80,150), (60,50), and (20,150), respectively. Filling in with a "smooth" curve, we get the picture as follows:

16

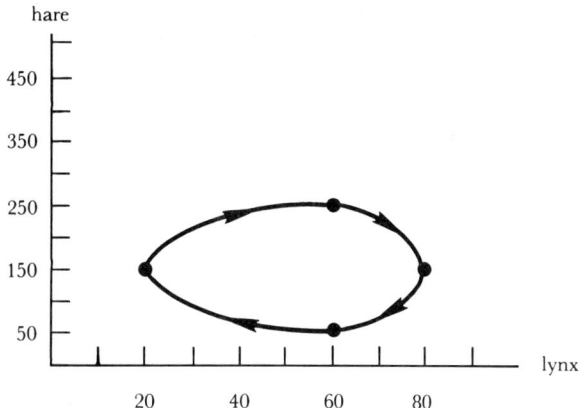

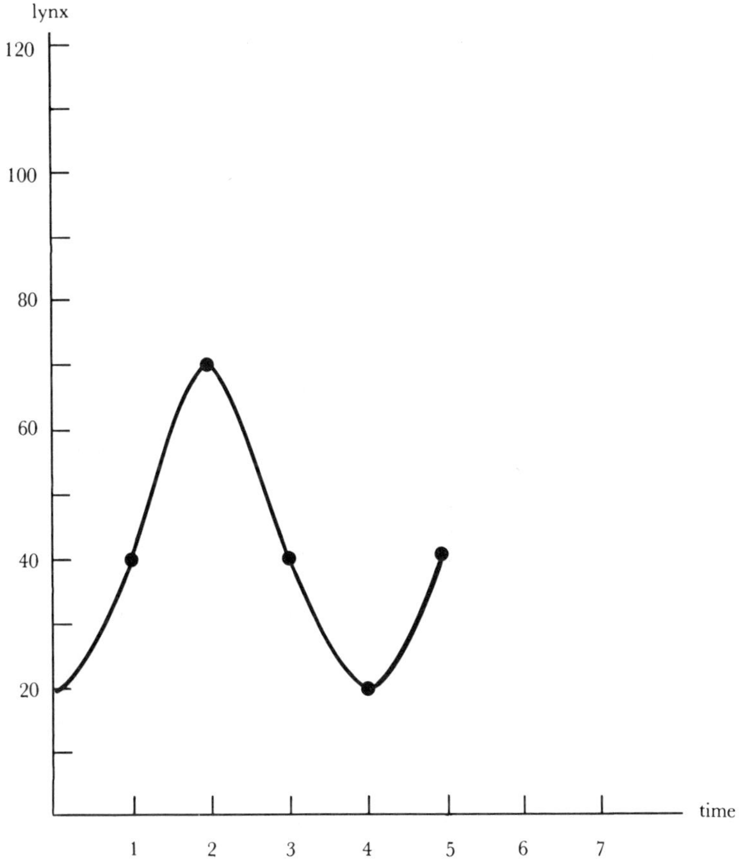

6. Since we are concerned with only the lynx we shall read off only the first coordinates of the points $P1$, $P2$, $P3$, and $P4$ (those that correspond to the lynx in the (L, H) pair). We get 20, 40, 70, and 40 for $P1$, $P2$, $P3$, and $P4$, respectively, which correspond to the times $t = 0$, 1, 2, and 3, respectively. Since the first coordinate of the population pair represents the number of lynx, we read only the first coordinate of the points $P1$, $P2$, $P3$, and $P4$ as those corresponding to times $t = 0, 1, 2$, and 3 (with some time unit attached). We then fill in with a curve motivated by the results of the Lotka-Volterra model.

4. References

Eisen, Martin. 1981. Graphical analysis of some difference equations in biology. UMAP Modules in Undergraduate Mathematics and Its Applications: Module 553. Arlington, MA: COMAP, Inc.

Uses only precalculus mathematics to consider a discrete model of the predator-prey interaction.

Elton, Charles, and Mary Nicholson. 1942. The ten-year cycle in numbers of the lynx in Canada. *Journal of Animal Ecology* 11:215–244.

This nonmathematical paper presents material from the archives of the Hudson's Bay Company. It contains fur returns over approximately 100 years, starting in 1821. The data are presented in graphical form that shows the great cyclical regularity and large (but varying) amplitude in a 10-year cycle. The relation of lynx to snowshoe hare is discussed.

Frauenthal, James C. *Introduction to Population Modeling*. The UMAP Expository Monograph Series. Boston: Birkhäuser.

Giordano, Frank R., and Stanley C. Leja. 1983. Competitive hunter models. UMAP Modules in Undergraduate Mathematics and Its Applications: Module 628. Arlington, MA: COMAP, Inc.

What if the two species are not predator and prey but both compete for the same food source? This module uses elementary concepts from calculus to examine this situation.

Lotka, A.J. 1925. *Elements of Physical Biology*. Baltimore, MD: Williams and Wilkins.

May, Robert M., editor. 1976. *Theoretical Ecology*. New York: Saunders.

This text contains articles that are much more sophisticated mathematically than the present paper. Chapter 6, written by M.P. Hassell, might be perused for its discussion of examples and the ideas of handling time of a prey by a predator and the probability of discovery of a prey by a predator. Hassell uses a discrete, rather than continuous, time variable.

Pielou, E.C. 1969. *An Introduction to Mathematical Ecology*. New York: Wiley.

Chapter 6 contains material on predation, much of which is difficult to follow without some guidance through the notation of differential equations. However, pp. 74–75 contain some interesting remarks concerning spatial heterogeneity, age distribution, and other nonuniformities, as well as some insights into the general process of mathematical modeling.

Wilson, Edward O., and William H. Bossert. 1971. *A Primer of Population Biology*. Sunderland, MA: Sinauer Associates.

Very useful information concerning the quantitative approach to biology and the formation of mathematical models is contained in Chapter 1. Although calculus notation is used, there is a helpful, informal, and intuitive explanation of such notation. Chapter 3 deals with predation and explains the Lotka-Volterra equations in much more depth than the present paper.

Young, Louise B. 1968. *Population in Perspective*. New York: Oxford University Press.

This is a collection of nonmathematical articles, starting with Malthus's seminal work on human population growth, and including Margaret Sanger on birth control and Sally Corrighar's research on crowding and stress.

About the Author

James Morrow studied mathematics at Miami University and at Florida State University, where he obtained his Ph.D. in 1976. His interests are currently focused on the ways that people learn mathematics. He and Charlene Morrow direct Mount Holyoke College's program for high school-age women, SummerMath.

UMAP Module 676

Modules in Undergraduate Mathematics and its Applications

Published in cooperation with the Society for Industrial and Applied Mathematics, the Mathematical Association of America, the National Council of Teachers of Mathematics, the American Mathematical Association of Two-Year Colleges, The Institute of Management Sciences, and the American Statistical Association.

Compartment Models in Biology

Ron Barnes

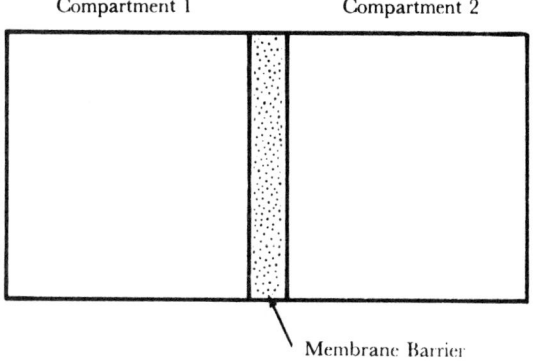

INTERMODULAR DESCRIPTION SHEET:	UMAP Unit 676
TITLE:	COMPARTMENT MODELS IN BIOLOGY
AUTHOR:	Ron Barnes Dept. of Applied Mathematical Sciences University of Houston, Downtown Houston, TX 77002
MATHEMATICAL FIELD:	Differential equations
APPLICATION FIELD:	Biology and biochemistry
TARGET AUDIENCE:	Students in a course in ordinary differential equations
ABSTRACT:	This module introduces compartment models and their applications.
PREREQUISITES:	Familiarity with systems of ordinary differential equations (but no linear algebra is required).

© Copyright 1987 by COMAP, Inc. All rights reserved.

COMAP, 60 Lowell Street, Arlington, MA 02174 (617) 641-2600

Compartment Models in Biology

Ron Barnes
Department of Applied Mathematical Sciences
University of Houston, Downtown
Houston, TX 77002

Table of Contents

1. INTRODUCTION 1
2. SOME GENERAL COMMENTS ON MODELING 1
3. COMPARTMENT ANALYSIS MODELS I 1
   3.1 Introduction 1
   3.2 Definitions and Descriptions 2
4. EXAMPLES OF COMPARTMENTAL MODELS 2
   4.1 Example 1—Simplest Non-Trivial Model 2
   4.2 Solution of Example 1 3
   4.3 Example 2—A More General Model and Solution . 5
5. COMPARTMENT ANALYSIS MODELS II 6
   5.1 General Comments 6
   5.2 A Generalized Compartmental Model 7
6. COMPARTMENTAL MATRICES AND INVERTIBILITY—
   AN EXAMPLE 9
7. EXERCISES 10
8. ANSWERS TO EXERCISES 16
9. SAMPLE EXAM 21
10. ANSWERS TO SAMPLE EXAM 22
11. REFERENCES 23

MODULES AND MONOGRAPHS IN UNDERGRADUATE
MATHEMATICS AND ITS APPLICATIONS (UMAP) PROJECT

The goal of UMAP was to develop, through a community of users and developers, a system of instructional modules in undergraduate mathematics and its applications to be used to supplement existing courses and from which complete courses may eventually be built.

The Project was guided by a National Advisory Board of mathematicians, scientists, and educators. UMAP was funded by a grant from the National Science Foundation and is now supported by the Consortium for Mathematics and Its Applications (COMAP), Inc., a nonprofit corporation engaged in research and development in mathematics education.

COMAP Staff

Solomon A. Garfunkel	Executive Director, COMAP
Laurie W. Aragon	Business Development Manager
Philip A. McGaw	Production Manager
Nancy Hawley	Copy Editor
Annemarie S. Morgan	Administrative Assistant

1. Introduction

The purpose of this module is to introduce the students of the biological and health sciences to some mathematical modeling techniques used in the biological sciences, but not widely accessible to undergraduate students even though the mathematics prerequisites are not very formidable. The models to be considered involve differential equations, so familiarity with systems of ordinary differential equations as treated in a first course in differential equations is required. Although a matrix is exhibited, no linear algebra is required. No specific training in biology or chemistry, beyond high school or first-year college level, is necessary.

The module begins with some general remarks about mathematical models and then uses examples to introduce the technique of compartmental analysis. Compartmental matrices are introduced, and a recent application involving these matrices is noted. Throughout the presentation, the module points out a number of additional avenues of investigation, with references that the interested reader may wish to pursue.

2. Some General Comments on Modeling

Modeling may be thought of as a tool for investigating real systems. A real system is observed, data are gathered, and the relationships observed in the real system are abstracted into a mathematical model. Investigation and analysis of this model lead to conjectures and predictions, which can then be compared to the behavior of the real system. This in turn leads to an improved model, and the iterative process continues until (we hope) a satisfactory model is obtained to explain the real system.

Models are often used because they (1) simplify the real system, (2) allow the testing of specific hypotheses that might endanger the real system, and (3) may help the investigator to recognize relationships in the real system that are masked by complexities and interactions.

3. Compartment Analysis Models I

3.1 Introduction

Biological organisms often are composed of natural physical components or compartments. For example, the human body contains the heart, lungs, and kidneys, each of which can be considered as a separate compartment. Compartments need not be limited to organs. We could analyze one substance by treating each of its different forms as a distinct compartment. For example, in the blood, red blood cells, plasma, and white cells could be treated as separate compartments. Thus, compartments may denote physical organs, different forms of a substance, or even different locations within an organism.

3.2 Definitions and Descriptions

A *compartmental system* is composed of a number of regions, called *components* or *compartments*. Each of these components or compartments is assumed to be a well-mixed, homogeneous unit. Individual compartments may interact with each other (and possibly their environments) by exchanging or transporting mass, momentum, and/or energy. *Compartmental analysis* deals with the mathematical analysis of the behavior of compartmental systems.

As time increases, the system may approach a *steady-state condition*. A steady-state condition is differentiated from an *equilibrium* state; equilibrium requires at least two compartments connected in such a way that exchange of mass and/or energy can take place between compartments in both directions, while the concept of steady state can apply to a single compartment.

To better understand the concept of compartments, consider a chamber divided into two regions (compartments) by a membrane. Water and a solute (dissolved materials) are present in each region. The membrane is assumed to be somewhat permeable to the solute S. We also assume the solutions on both sides of the membrane are *well stirred*, i.e., the instant any material S is added to one solution, it is immediately distributed throughout that compartment, but not so quickly across the membrane to the other compartment. In other words, we assume that the equilibrium with respect to mass transport is rapid within the solution, compared to across the membrane.

4. Examples of Compartmental Models

4.1 Example 1—Simplest Non-Trivial Model

Through the barrier, a solute (e.g., a trace chemical) can diffuse from one region to the other at a rate proportional to the difference

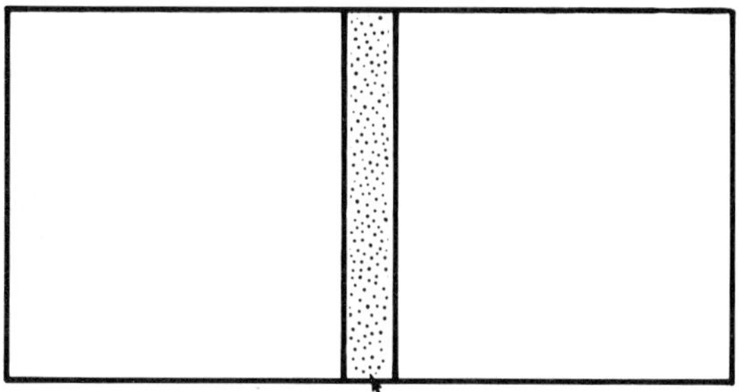

Figure 1. Simplest compartmental model.

$c_1 - c_2$ in concentration in the two compartments (from the higher to the lower concentration). We wish to find the concentration $c_i(t)$ in each compartment i at time t.

4.2 Solution of Example 1

We first note that

$$y_1(t) = c_1(t)v_1 \quad \text{and} \quad y_2(t) = c_2(t)v_2, \tag{1}$$

where y_i is the amount of solution S in compartment i at time t; $c_i(t)$ is the concentration of S in compartment i at time t; and v_i is the volume of compartment i. We denote derivatives with respect to time as $\dot{y}_i(t)$ and $\ddot{y}_i(t)$. Since the change in the amount of solute in compartment i is proportional to the difference in concentration in the two compartments, it is clear that

$$\dot{y}_1(t) = k(c_2(t) - c_1(t)) \quad \text{and} \quad \dot{y}_2(t) = k(c_1(t) - c_2(t)) \tag{2}$$

for some constant $k > 0$.

We can differentiate equations (1) and then solve for $c_i(t)$. If we do this, we see that

$$\dot{c}_1(t) = \frac{\dot{y}_1(t)}{v_1} = \frac{k(c_2(t) - c_1(t))}{v_1}; \tag{3}$$

and

$$\dot{c}_2(t) = \frac{\dot{y}_2(t)}{v_2} = \frac{k(c_1(t) - c_2(t))}{v_2}$$

$$= \frac{-\dot{y}_1(t)}{v_2} = \frac{-\dot{c}_1(t)v_1}{v_2} = \frac{-v_1}{v_2}\dot{c}_1(t). \tag{4}$$

Now, by differentiating (3) and substituting (4), we see that

$$\ddot{c}_1(t) = \frac{k[\dot{c}_2(t) - \dot{c}_1(t)]}{v_1} = \frac{k}{v_1}\left[-\frac{v_1}{v_2}\dot{c}_1(t) - \dot{c}_1(t)\right],$$

or

$$\ddot{c}_1(t) = -k\left[\frac{1}{v_2} + \frac{1}{v_1}\right]\dot{c}_1(t),$$

or

$$\ddot{c}_1(t) + k\left[\frac{1}{v_1} + \frac{1}{v_2}\right]\dot{c}_1(t) = 0. \tag{5}$$

Equation (5) is a simple second-order differential equation with constant coefficients. To solve it, initial conditions $c_1(t_0)$ and $\dot{c}_1(t_0)$ need to be given for some time t_0. The initial concentrations $c_1(t_0)$ and $c_2(t_0)$ are specified by the investigator, and the diffusion rate $\dot{c}_1(t)$ can be found using equation (5). If $t_0 = 0$, then the solution is given by

$$c_1(t) = c_1(0) - \frac{v_2}{v_1 + v_2}[c_2(0) - c_1(0)]$$

$$\times \left\{\exp\left[-k\left(\frac{1}{v_1} + \frac{1}{v_2}\right)t\right] - 1\right\}. \tag{6}$$

Similarly, we have

$$c_2(t) = c_2(0) - \frac{v_1}{v_1 + v_2}[c_1(0) - c_2(0)]$$

$$\times \left\{\exp\left[-k\left(\frac{1}{v_1} + \frac{1}{v_2}\right)t\right] - 1\right\}. \tag{7}$$

If we let $t \to \infty$, we see that the two compartments tend to the same concentration

$$c_1(\infty) = c_2(\infty) = \frac{v_1 c_1(0) + v_2 c_2(0)}{v_1 + v_2}. \tag{8}$$

Notice that it is also true that $c_1(t)v_1 + c_2(t)v_2 = c_1(0)v_1 + c_2(0)v_2$ for any value of t.

4.3 Example 2—A More General Model and Solution

A more general case for the two-compartment model allows for the possibility of different rates of permeability, k_{12} from compartment 1 to 2 and k_{21} from compartment 2 to 1, where these values replace the single value k of **Example 1**. While equations (1) remain the same for the new example, equations (2) are replaced by

$$\dot{y}_1(t) = k_{21}c_2(t) - k_{12}c_1(t) \quad \text{and} \quad \dot{y}_2(t) = k_{12}c_1(t) - k_{21}c_2(t). \tag{9}$$

From (9) it follows that

$$\dot{y}_1(t) + \dot{y}_2(t) \equiv 0. \tag{10}$$

Equation (10) is called the *mass balance equation*. It simply reflects conservation of mass of the solute, since any loss from one compartment must be gained by the other compartment. We do not get any corresponding *concentration* balance equation, even in the simpler case of **Example 1**; so $\dot{c}_1(t) \neq -\dot{c}_2(t)$ unless $v_1 = v_2$.

By using (1) and (9), one can assume initial conditions $c_1(0)$ and $c_2(0)$, proceed as in **Example 1**, and obtain the following results:

$$c_1(t) = c_1(0) - \frac{v_2(k_{21}c_2(0) - k_{12}c_1(0))}{k_{21}v_1 + k_{12}v_2}$$

$$\times \left[\exp\left[-\left(\frac{k_{12}}{v_1} + \frac{k_{21}}{v_2} \right) t \right] - 1 \right] \tag{11}$$

and

$$c_2(t) = c_2(0) - \frac{v_1(k_{12}c_1(0) - k_{21}c_2(0))}{k_{21}v_1 + k_{12}v_2}$$

$$\times \left[\exp\left[-\left(\frac{k_{12}}{v_1} + \frac{k_{21}}{v_2} \right) t \right] - 1 \right]. \tag{12}$$

In the limit as $t \to \infty$, we get

$$c_1(\infty) = \frac{k_{21}(c_1(0)v_1 + c_2(0)v_2)}{k_{21}v_1 + k_{12}v_2}$$

and

$$c_2(\infty) = \frac{k_{12}(c_1(0)v_1 + c_2(0)v_2)}{k_{21}v_1 + k_{12}v_2}. \tag{13}$$

In this case, $c_1(\infty) \neq c_2(\infty)$ in general, but

$$c_1(\infty)v_1 + c_2(\infty)v_2 = c_1(0)v_1 + c_2(0)v_2 \tag{14}$$

and

$$c_1(t)v_1 + c_2(t)v_2 = c_1(0)v_1 + c_2(0)v_2. \tag{15}$$

5. Compartment Analysis Models II

5.1 General Comments

In elementary compartmental analysis, three general assumptions are usually made:

i. constant-volume compartments,

ii. well-mixed compartments,

iii. constant transport coefficient(s).

For a given compartmental system, solution then becomes a matter of setting up the differential equations, for the rate of transport of a substance S with respect to each compartment, and solving the resulting system.

In both of the examples given thus far, the rate of change of the substance S within a compartment depended only on the concentration of S in that and the neighboring compartments; so, e.g., we had in **Example 1**, $\dot{y}_1(t) = k(c_2(t) - c_1(t))$. In many cases, the relationship is not so simple. The rate of removal of S from a compartment may depend on other factors. For example, the rate of oxygen removal from the blood by tissue is more dependent on metabolism than on the oxygen pressure in the blood.

In the general case, the change in the amount of material S in a compartment may be due to flow through the compartment or to the creation and/or utilization of the material within the compartment. In this general case, we can express the transport of a material S in a compartment i by

$$y_i = I_s - L_s + U_s, \tag{16}$$

where

y_i = change of material S in compartment i;

I_s = input of material S;

$$L_s = \text{loss of material } S \text{ (including both concentration-dependent and independent terms);}$$

$$U_s = \text{utilization of } S \text{ (with both concentration-dependent and independent terms), where utilization may involve both creation and destruction of the } S \text{ material.}$$

The forms of I, L, and U are determined by the specific model.

Finally, a complex system may involve transport of more than one substance S, so the system of equations may include equations for transport of a number of different substances or species.

We now note some considerations of how the process of compartmental analysis is applied when one or more of the assumptions is not valid. While the three assumptions permit ideal conditions, the real system under consideration often will not correspond to either perfectly mixed or constant-volume compartmental models.

5.2 A Generalized Compartmental Model

The general form of the mass transport relationship is often given in the form (for compartment k)

$$\frac{dy_k}{dt} = \frac{d(\rho_k V_k)}{dt} = \sum_i \rho_{ik} Q_{ik} - \sum_j \rho_{kj} Q_{kj} + \sum_h D_{hk}, \qquad (17)$$

where

y_k = mass of S in compartment k;

V_k = volume of component k;

$\rho_k = \dfrac{y_k}{V_k}$ = mass density in compartment k;

ρ_{ij} = mass density of material flowing from i to j;

Q_{ij} = volumetric flow rate of material from i to j;

D_{ij} = net rate of diffusion of mass from i to j.

This equation can be expressed in words by

rate of mass accumulation in the kth compartment	=	net rate of mass entering the compartment by convection (volumetric flow)	+	net rate of mass entering by diffusion

If the well-mixing assumption is not valid, then within any compartment the mass density $\rho_k = \rho_k(x, y, z, t)$ is spatially dependent. In such a case, the system reduces to a *partial* differential equations model. See [Banks 1979] for further discussion of this case.

When the well-mixed assumption is accepted, and in addition one assumes that the mass density of material flowing out of a compartment is the same as the density within the compartment, i.e., $\rho_{kj} = \rho_k$, then the form of (17) becomes

$$\frac{d(\rho_k V_k)}{dt} = \sum_i \rho_i Q_{ik} - \sum_j \rho_k Q_{kj} + \sum_h D_{hk} \qquad (18)$$

for compartment k.

One can write species mass transport equations analogous to (17) for species α in compartment k as

$$\frac{dy_k^\alpha}{dt} = \dot{y}_k^\alpha = \sum_i \rho_{ik}^\alpha Q_{ik} - \sum_j \rho_{kj}^\alpha Q_{kj} + \sum_h D_{hk}^\alpha + V_k r_k^\alpha, \qquad (19)$$

where r_k^α is the net rate of production of species α by reaction with compartment k, per unit volume.

By definition,

$$y_k = \sum_\alpha y_k^\alpha, \quad \rho_{ij} = \sum_\alpha \rho_{ij}^\alpha, \quad D_{ij} = \sum_\alpha D_{ij}^\alpha. \qquad (20)$$

Then, by summing (19) over α and using (20), we see by comparison with (17) that

$$\sum_\alpha V_k r_k^\alpha = V_k \sum_\alpha r_k^\alpha = 0 \Rightarrow \sum r_k^\alpha = 0 \qquad (21)$$

This final expression demonstrates the principle that mass is not created or destroyed by the interaction, but merely converted from one species to another.

Similarly, for a perfectly mixed compartmental system, a version of (18) is available, namely

$$\frac{d}{dt}(\rho_k^\alpha V_k) = \sum_i \rho_i^\alpha Q_{ik} - \sum_j \rho_k^\alpha Q_{kj} + \sum_h D_{hk}^\alpha + V_k r_k^\alpha. \qquad (22)$$

In general, to solve systems of equations of the form (22), one needs to know more about the relationship between D_{hk}^α and ρ_j^α. Banks discusses this at length [1979]. In addition, he considers examples of systems where volumes of the compartments do not remain constant. For a deeper discussion, the interested reader should consult [Banks and Palatt 1975, Chapter IV].

The standard approach to solving general systems of linear ordinary differential equations, [Finizio and Ladas 1978, 56–58, 162–163] uses matrix methods. Compartmental systems have a number of special properties that lead to some rather interesting results. To demonstrate these, we outline some ideas from a paper by David Anderson [1979].

6. Compartmental Matrices and Invertibility—An Example

In the modeling and simulation of phenomena in biology, chemistry, and medicine, a special class of square matrices arises with the following properties:

    i. each off-diagonal entry is non-negative;

    ii. each diagonal element is negative;

    iii. all columns but one add to zero;

    iv. the exceptional column sums to a negative value.

As an example, we note

$$K = \begin{bmatrix} -3 & 0 & 0 \\ 1 & -4 & 2 \\ 1 & 4 & -2 \end{bmatrix}.$$

Such matrices occur in connection with compartmental models associated with radioactive tracers in the human body. Consider the case of three compartments, where exit to the outside environment is only through the first compartment. In this case, it is reasonable to assume that the change in the amount of tracer dye in a compartment is proportional to the amount of dye entering and leaving the compartment from other compartments, plus any input function b_i that might inject dye into the various components. The k_{ij} fractional transfer coefficients describe the relative ease with which dye can move from compartment j to compartment i. A transfer coefficient of 0 means no transfer is possible, while a coefficient of 1 means that all of the dye may pass from j to i.

Denoting by $x_i(t)$ the amount of a tracer dye in the ith compartment, and assuming a linear model, we obtain the system model

$$\dot{x}_1 = (-k_{21} - k_{31} - k_{01})x_1 + k_{12}x_2 + k_{13}x_3 + b_1(t);$$
$$\dot{x}_2 = k_{21}x_1 + (-k_{12} - k_{32})x_2 + k_{23}x_3 + b_2(t);$$
$$\dot{x}_3 = k_{31}x_1 + k_{32}x_2 + (-k_{13} - k_{23})x_3 + b_3(t),$$

where k_{ij} is the non-negative fractional transfer coefficient from compartment j to compartment i; k_{01} is the positive exit-rate constant; and $b_i(t)$ denote the input functions.

The system can then be written in matrix form as

$$\dot{x} = kx + b,$$

with

$$k = \begin{bmatrix} -(k_{21} + k_{31} + k_{01}) & k_{12} & k_{13} \\ k_{21} & -(k_{12} + k_{32}) & k_{23} \\ k_{31} & k_{32} & -(k_{13} + k_{23}) \end{bmatrix}.$$

Observe that the assumptions are satisfied.

A *compartmental matrix* with n states is an $n \times n$ matrix $k = [k_{ij}]$, with $k_{ij} \geq 0$ if $i \neq j$ and $k_{jj} = -\Sigma_{i \neq j}k_{ij} - k_{0j}$ when $k_{0j} \geq 0$. We assume $k_{jj} < 0$ for all j (to be non-trivial). If, in addition, we require $k_{0j} = 0$ for all j, except for one value $j = m$ for which $k_{0m} \geq 0$ (as in our example above), then we call such a matrix k a *single-exit compartmental* (SEC) matrix.

Solving the system of differential equations for a single-exit compartmental model reduces to solving a linear system of algebraic equations with an SEC matrix as the matrix of coefficients. Anderson [1979] develops simple sufficient conditions for the invertibility of an SEC matrix, discusses properties of the inverse matrix, and develops an interactive procedure that converges to the solution when one exists. This procedure is superior to Gaussian elimination, Crout reduction with interactive improvement, and singular value decomposition for the special class of SEC matrices [Farnsworth 1978].

7. Exercises

1. Solve equation (5) and show that the solution is given by equation (6).

remain constant. For a deeper discussion, the interested reader should consult [Banks and Palatt 1975, Chapter IV].

The standard approach to solving general systems of linear ordinary differential equations, [Finizio and Ladas 1978, 56–58, 162–163] uses matrix methods. Compartmental systems have a number of special properties that lead to some rather interesting results. To demonstrate these, we outline some ideas from a paper by David Anderson [1979].

6. Compartmental Matrices and Invertibility—An Example

In the modeling and simulation of phenomena in biology, chemistry, and medicine, a special class of square matrices arises with the following properties:

    i. each off-diagonal entry is non-negative;

    ii. each diagonal element is negative;

    iii. all columns but one add to zero;

    iv. the exceptional column sums to a negative value.

As an example, we note

$$K = \begin{bmatrix} -3 & 0 & 0 \\ 1 & -4 & 2 \\ 1 & 4 & -2 \end{bmatrix}.$$

Such matrices occur in connection with compartmental models associated with radioactive tracers in the human body. Consider the case of three compartments, where exit to the outside environment is only through the first compartment. In this case, it is reasonable to assume that the change in the amount of tracer dye in a compartment is proportional to the amount of dye entering and leaving the compartment from other compartments, plus any input function b_i that might inject dye into the various components. The k_{ij} fractional transfer coefficients describe the relative ease with which dye can move from compartment j to compartment i. A transfer coefficient of 0 means no transfer is possible, while a coefficient of 1 means that all of the dye may pass from j to i.

Denoting by $x_i(t)$ the amount of a tracer dye in the ith compartment, and assuming a linear model, we obtain the system model

$$\dot{x}_1 = (-k_{21} - k_{31} - k_{01})x_1 + k_{12}x_2 + k_{13}x_3 + b_1(t);$$
$$\dot{x}_2 = k_{21}x_1 + (-k_{12} - k_{32})x_2 + k_{23}x_3 + b_2(t);$$
$$\dot{x}_3 = k_{31}x_1 + k_{32}x_2 + (-k_{13} - k_{23})x_3 + b_3(t),$$

where k_{ij} is the non-negative fractional transfer coefficient from compartment j to compartment i; k_{01} is the positive exit-rate constant; and $b_i(t)$ denote the input functions.

The system can then be written in matrix form as

$$\dot{x} = kx + b,$$

with

$$k = \begin{bmatrix} -(k_{21} + k_{31} + k_{01}) & k_{12} & k_{13} \\ k_{21} & -(k_{12} + k_{32}) & k_{23} \\ k_{31} & k_{32} & -(k_{13} + k_{23}) \end{bmatrix}.$$

Observe that the assumptions are satisfied.

A *compartmental matrix* with n states is an $n \times n$ matrix $k = [k_{ij}]$, with $k_{ij} \geq 0$ if $i \neq j$ and $k_{jj} = -\Sigma_{i \neq j} k_{ij} - k_{0j}$ when $k_{0j} \geq 0$. We assume $k_{jj} < 0$ for all j (to be non-trivial). If, in addition, we require $k_{0j} = 0$ for all j, except for one value $j = m$ for which $k_{0m} \geq 0$ (as in our example above), then we call such a matrix k a *single-exit compartmental* (SEC) matrix.

Solving the system of differential equations for a single-exit compartmental model reduces to solving a linear system of algebraic equations with an SEC matrix as the matrix of coefficients. Anderson [1979] develops simple sufficient conditions for the invertibility of an SEC matrix, discusses properties of the inverse matrix, and develops an interactive procedure that converges to the solution when one exists. This procedure is superior to Gaussian elimination, Crout reduction with interactive improvement, and singular value decomposition for the special class of SEC matrices [Farnsworth 1978].

7. Exercises

1. Solve equation (5) and show that the solution is given by equation (6).

2. Use equations (6) and (7) to show that (8) is true.

3. Explain why you should expect the equilibrium concentrations $c_1(\infty)$ and $c_2(\infty)$ to be equal.

4. Use equations (1) and (9) to obtain the differential equation analogous to (5) that you would need to solve to obtain equation (11).

5. Use equation (12) to show that the limiting concentration for **Example 2** of $c_2(t)$ is given by (13).

6. Explain why equations (14) and (15) are true, though in general we have $c_1(\infty) \neq c_2(\infty)$.

7. In section 5.2, it was noted that if the well-mixing assumption is violated, then the result is a system of *partial* differential equations. Why is this the case?

8. Are the matrices below compartmental matrices? SEC matrices?

   a. $\begin{bmatrix} -0.3 & 0.1 & 0.0 & 0.2 \\ 0.1 & -0.1 & 0.2 & 0.2 \\ 0.2 & 0.0 & -0.4 & 0.1 \\ 0.0 & 0.0 & 0.0 & -0.5 \end{bmatrix}$

   b. $\begin{bmatrix} -2 & 1 & 0 & 0 \\ 1 & -1 & -1 & 0 \\ 0 & 0 & 1 & 0 \\ 0 & 0 & 0 & -1 \end{bmatrix}$

9. In studying liver disease, Anderson et. al. [1977] considered a test measuring the disappearance of radioactive Rose Bengal. Their study led to a four-compartment mathematical model of tracer transport.

   The study was motivated by the transport of a substance bilirubin, which forms the basis for many tests used in determining the nature of certain blood and liver disorders. An excess of bilirubin in the blood and body tissue causes jaundice, the yellow discoloration associated with liver diseases. Recent studies suggested to the authors that the idealized pathway of radioactive Rose Bengal through the biliary system is given in **Figure 2**. If $X_i(t)$ denotes the amount of the tracer in any component i at time t, then the rate of change of the amount $X_i(t)$, denoted by $\dot{X}_i(t)$, is the linear input minus output of that compartment.

   a. Denote by k_{ij} the constant fractional transfer coefficient from compartment j to compartment i.

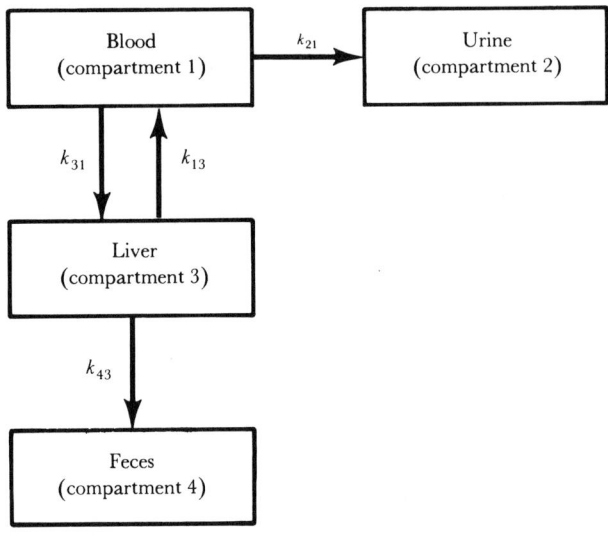

Figure 2.

>**b.** Set up the corresponding system of differential equations that describes this compartmental model. You can denote the initial conditions $X_i(0)$ by X_{i_0}.

The authors note that classical compartmental analysis has limited success in this setting, because of the small amount of data.

> In a reasonable period of time (say 72 hours) it may be possible to obtain only three or four urine or fecal specimens. Another related shortcoming of classical compartmental techniques is the short interval of time required between blood samples, and thus the large number of samples that must be taken and analyzed—making the technique not very appealing as a clinical procedure.

The data are needed to estimate the k_{ij} parameters.

The authors solved these difficulties with a discretized model. They reduced the problem to a four-state Markov chain with states b = blood; u = urine; l = liver; and f = feces, where urine and feces are absorbing states (**Figure 3**). The transition matrix (P_{ij}) for the Markov chain is related to the fractional transfer coefficient matrix (k_{ij}). In particular, P_{ij} is the proportion of Rose Bengal in the ith compartment that is transferred to the jth compartment in one hour. Under certain conditions, the authors show that $P_{ij} = hk_{ji} + \delta_{ij}$, where h is a small non-negative constant.

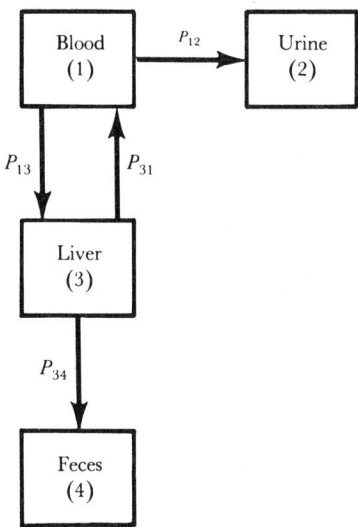

Figure 3.

Utilizing this Markov chain reformulation, they were able to estimate the P_{ij} parameters and construct the graphs for the $X_i(t)$ distributions for the percentage of Rose Bengal in the blood, urine, and feces over a 72-hour period (**Figure 4**).

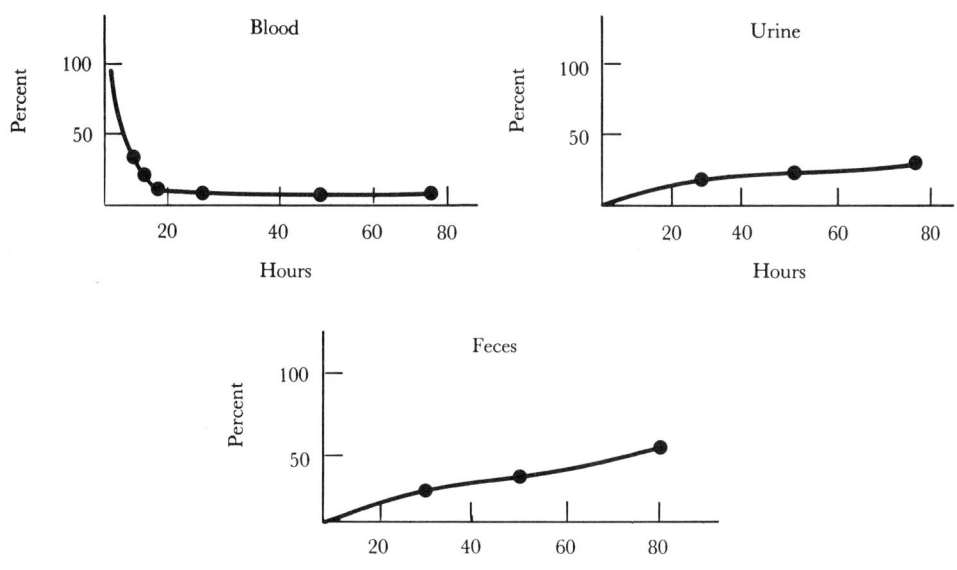

Figure 4.

10. The *cell cycle* is the path that a cell travels to get to the stage when it is ready to replicate (by mitosis) into two new cells. The first phase is the G_1 phase, in which the cell begins to synthesize RNA and consequently protein. The G_1 phase ends with the beginning of DNA replication, which constitutes the S phase. RNA and protein synthesis continue throughout this phase, but at a reduced rate. With the end of the DNA replication comes the G_2 phase. This phase is the preparation for mitosis (cell division). During this phase RNA and protein synthesis continue at increased rates as the cell prepares to divide. Mitosis, the actual subdivision and production of two new cells, is the M phase. In the following discussion, the G_2 and M phases are combined as the $G_2 + M$ phase.

Dosik et. al. [1981] developed a method for examining the cell-cycle dynamics. One of their studies involved Chinese hamster ovary cells that were continuously exposed to a substance, calcemid, for blocking to suppress reentry into the G_1 phase. Changes in the cell-cycle phases (compartments) distribution were monitored by DNA flow cytometry. Analysis of the rate of $G_2 + M$ compartment accumulation after the addition of colcemid then permitted calculation of a number of cycle transit parameters, including the complete cell-cycle transit time $T_c = T_1 + T_2 + T_3$, where

T_1 = the phase transit time in G_1;
T_2 = the phase transit time in S;
T_3 = the phase transit time in $G_2 + M$.

The authors showed that these cell-cycle transit times depend on the proportion or percentage $I_i(t)$ of cells in each of the different phases (compartment).

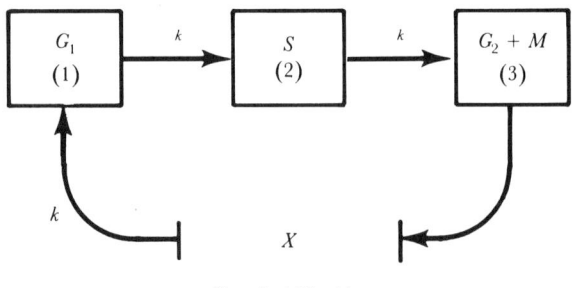

Figure 5.

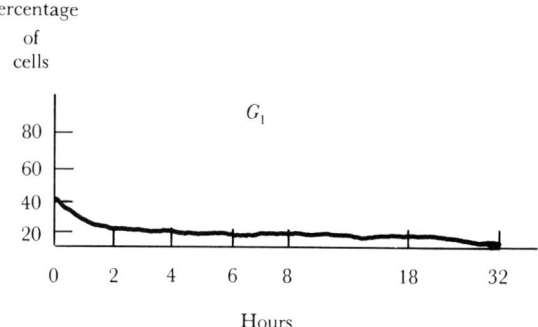

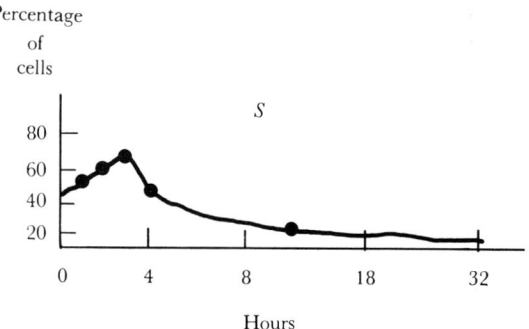

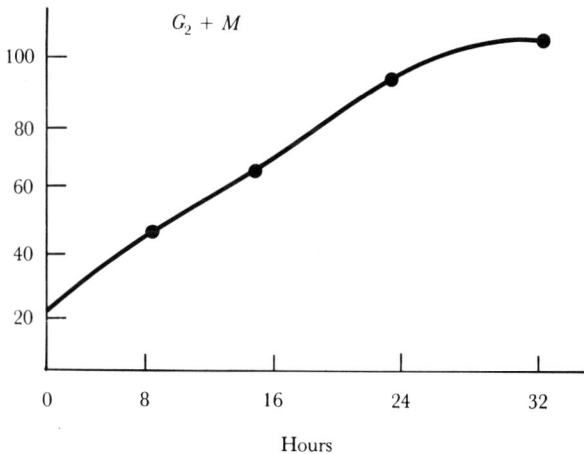

Figure 6.

Their paper sought to determine the quantities $I_i(t)$ when an exponential source of cells is supplied to G_1 and the cells at stage $G_2 + M$ are blocked to prevent reentry into the G_1 phase (**Figure 5**). Note that k acts like a fractional transfer coefficient and is assumed constant for the different phases.

a. Set up the appropriate system of differential equations.
b. Solve the system using the following boundary conditions:

$$I_1(0) = 2(1 - e^{-kT_1});$$

$$I_2(0) = 2e^{-kT_1}(1 - e^{-kT_2});$$

$$I_3(0) = 2e^{-k(T_1+T_2)}(1 - e^{-kT_3}).$$

Using data, the authors constructed the graphs for the percentage of cells in each phase $I_i(t)$ (**Figure 6**). The authors show how k can be estimated from the data and how the transit parameters can be estimated. They note that their estimates compare favorably with those used in other techniques to deal with similar data.

8. Answers to Exercises

1. To solve $\ddot{c}_1 + k\left[\dfrac{1}{v_1} + \dfrac{1}{v_2}\right]\dot{c}_1 = 0$, we note that the general solution for such an equation with constant coefficients is given by

$$c_1 = a_1 \exp[\lambda_1 t] + a_2 \exp[\lambda_2 t],$$

where λ_1 and λ_2 are the eigenvalues of

$$\lambda^2 + k\left(\dfrac{1}{v_1} + \dfrac{1}{v_2}\right)\lambda = 0.$$

Since the left-hand side factors as

$$\lambda\left(\lambda + k\left(\dfrac{1}{v_1} + \dfrac{1}{v_2}\right)\right) = 0,$$

we see that the roots are $\lambda_1 = 0$, $\lambda_2 = -k\left(\dfrac{1}{v_1} + \dfrac{1}{v_2}\right)$. The general solution then becomes

$$c_1(t) = a_1 \exp[0t] + a_2 \exp\left[-k\left(\frac{1}{v_1} + \frac{1}{v_2}\right)t\right]$$

$$= a_1 + a_2 \exp\left[-k\left(\frac{1}{v_1} + \frac{1}{v_2}\right)t\right].$$

To get the values for a_1 and a_2, we note that since

$$\dot{c}_1(t) = \frac{k}{v_1}(c_2(t) - c_1(t)) \quad \text{we can use the fact that}$$

$$\dot{c}_1(0) = \frac{k}{v_1}(c_2(0) - c_1(0)) \quad \text{to solve for } a_1 \text{ and } a_2 \text{ as follows:}$$

$$c_1(0) = a_1 + a_2 \exp[0] = a_1 + a_2$$

$$\dot{c}_1(0) = -k\left(\frac{1}{v_1} + \frac{1}{v_2}\right) a_2 \exp[0] = -a_2 k\left(\frac{1}{v_1} + \frac{1}{v_2}\right).$$

Solving these 2 equations gives

$$a_2 = \frac{-\dot{c}_1(0)}{k\left(\frac{1}{v_1} + \frac{1}{v_2}\right)} = \frac{-k(c_2(0) - c_1(0))}{v_1 k\left(\frac{1}{v_1} + \frac{1}{v_2}\right)}.$$

Multiplying numerator and denominator by v_2 yields

$$a_2 = \frac{-v_2(c_2(0) - c_1(0))}{v_1 + v_2}.$$

Substituting this value for a_2 into the equation

$$a_1 + a_2 = c_1(0)$$

gives

$$a_1 = c_1(0) - \frac{(-v_2(c_2(0) - c_1(0)))}{v_1 + v_2}.$$

Finally, substituting these values into

$$c_1(t) = a_1 + a_2 \exp\left[-k\left(\frac{1}{v_1} + \frac{1}{v_2}\right)t\right]$$

gives

$$c_1(t) = c_1(0) - \frac{v_2}{v_1 + v_2}[c_2(0) - c_1(0)]$$

$$\times \left[\exp\left[-k\left(\frac{1}{v_1} + \frac{1}{v_2}\right)t\right] - 1\right].$$

2.

$$c_1(t) = c_1(0) - \frac{v_2}{v_1 + v_2}[c_2(0) - c_1(0)]$$

$$\times \left[\exp\left[-k\left(\frac{1}{v_1} + \frac{1}{v_2}\right)t\right] - 1\right].$$

$$c_2(t) = c_2(0) - \frac{v_1}{v_1 + v_2}[c_1(0) - c_2(0)]\left[\exp\left[-k\left(\frac{1}{v_1} + \frac{1}{v_2}\right)t\right] - 1\right].$$

If we let $t \to \infty$ in these equations, we get

$$c_1(\infty) = c_1(0) - \frac{v_2}{v_1 + v_2}[c_2(0) - c_1(0)][0 - 1];$$

$$c_2(\infty) = c_2(0) - \frac{v_1}{v_1 + v_2}[c_1(0) - c_2(0)][0 - 1].$$

These reduce to

$$c_1(\infty) = c_1(0) + \frac{v_2}{v_1 + v_2}(c_2(0) - c_1(0))$$

$$= \frac{c_1(0)v_1 + c_1(0)v_2 + v_2(c_2(0) - c_1(0))}{v_1 + v_2}$$

$$= \boxed{\frac{c_1(0)v_1 + c_2(0)v_2}{v_1 + v_2}};$$

and

$$c_2(\infty) = c_2(0) + \frac{v_1}{v_1 + v_2}(c_1(0) - c_2(0))$$

$$= \frac{c_2(0)v_1 + c_2(0)v_2 + v_1(c_1(0) - c_2(0))}{v_1 + v_2}$$

$$= \boxed{\frac{c_1(0)v_1 + c_2(0)v_2}{v_1 + v_2}}.$$

3. You would expect the concentrations to be constant at equilibrium. Unequal concentrations would result in the migration of S from the area of lower concentration, which would contradict the assumption of equilibrium.

4. $y_1(t) = c_1(t)v_1$, $y_2(t) = c_2(t)v_2$ and $\dot{y}_1(t) = k_{21}c_2(t) - k_{12}c_1(t)$, $\dot{y}_2(t) = k_{12}c_1(t) - k_{21}c_2(t)$. Therefore,

$$\dot{c}_1(t) = \frac{\dot{y}_1(t)}{v_1} = \frac{k_{21}c_2(t) - k_{12}c_1(t)}{v_1} \tag{23}$$

and

$$\dot{c}_2(t) = \frac{\dot{y}_2(t)}{v_2} = \frac{k_{12}c_1(t) - k_{21}c_2(t)}{v_2} = \frac{-v_1}{v_2}\dot{c}_1(t). \tag{24}$$

Next we differentiate (23) and substitute (24) to obtain

$$\ddot{c}_1(t) = \frac{k_{21}\dot{c}_2(t) - k_{12}\dot{c}_1(t)}{v_1}$$

$$= \frac{1}{v_1}\left(k_{21}\left(\frac{-v_1}{v_2}\dot{c}_1(t)\right) - k_{12}\dot{c}_1(t)\right)$$

$$= \frac{-k_{21}\dot{c}_1(t)}{v_2} - \frac{k_{12}\dot{c}_1(t)}{v_1}$$

$$= -\left(\frac{k_{21}}{v_2} + \frac{k_{12}}{v_1}\right)\dot{c}_1(t).$$

The resulting differential equation then becomes

$$\boxed{\ddot{c}_1(t) + \left(\frac{k_{12}}{v_1} + \frac{k_{21}}{v_2}\right)\dot{c}_1(t) = 0}.$$

5. Since

$$c_2(t) = c_2(0) - \frac{v_1(k_{12}c_1(0) - k_{21}c_2(0))}{k_{21}v_1 + k_{12}v_2}$$

$$\times \left[\exp\left[-\left(\frac{k_{12}}{v_1} + \frac{k_{21}}{v_2}\right)t\right] - 1\right],$$

we let $t \to \infty$ and get

$$c_2(\infty) = c_2(0) - \frac{v_1 k_{12}c_1(0) - v_1 k_{21}c_2(0)}{k_{21}v_1 + k_{12}v_2}[0 - 1],$$

or

$$c_2(\infty) = c_2(0) + \frac{v_1 k_{12}c_1(0) - v_1 k_{21}c_2(0)}{k_{21}v_1 + k_{12}v_2}$$

$$= \frac{k_{21}c_2(0)v_1 + k_{12}c_2(0)v_2 + v_1 k_{12}c_1(0) - v_1 k_{21}c_2(0)}{k_{21}v_1 + k_{12}v_2}$$

$$= \frac{k_{12}(c_1(0)v_1 + c_2(0)v_2)}{k_{21}v_1 + k_{12}v_2}.$$

It is similar for $c_1(\infty)$.

6. Equations (14) and (15) merely state that the total mass of the concentrate remains the same at equilibrium, initially or at any intermediate time t_0. In general, $c_1(\infty) \neq c_2(\infty)$ because the rates at which the solute enters and exits the two chambers are not the same.

7. In this case, we have

$$\frac{d(\rho_k v_k)}{dt} = \frac{d(\rho_k(x, y, z, t)v_k)}{dt}$$

$$= \left(\frac{\partial \rho_k}{\partial x}\frac{dx}{dt} + \frac{\partial \rho_k}{\partial y}\frac{dy}{dt} + \frac{\partial \rho_k}{\partial z}\frac{dz}{dt} + \frac{d\rho_k}{dt} \right) v_k.$$

8. **a.** $k_{ij} \geq 0$ for $i \neq j$ and $k_{jj} < 0$ for all j, so it is a compartmental matrix.
   b. $k_{33} \geq 0$, so this is not a compartmental matrix. Notice **a.** is an SEC matrix; only the third column sums to a negative value, therefore:

$$k_{33} = -\sum_{i \neq 3} k_{i3} - k_{03} \quad \text{or}$$

$$-.4 = -.2 - k_{03} \quad \text{so} \quad \boxed{k_{03} = .2}.$$

9.

$$\dot{X}_1 = -k_{21}X_1 - k_{31}X_1 + k_{13}X_3$$
$$\dot{X}_2 = +k_{21}X_1$$
$$\dot{X}_3 = k_{31}X_1 - k_{13}X_3 - k_{43}X_3$$
$$\dot{X}_4 = k_{43}X_3$$
$$X_i(0) = X_{i_0}$$

10. a.

$$\dot{I}_1 = -kI_1$$
$$\dot{I}_2 = kI_1 - kI_2$$
$$\dot{I}_3 = kI_2$$

b. The eigenvalues are $\lambda = 0$, $\lambda = -k$ and $\lambda = -k$, and the general solution can be expressed as

$$I(t) = \begin{bmatrix} I_1(t) \\ I_2(t) \\ I_3(t) \end{bmatrix} = c_1 \begin{bmatrix} 0 \\ 0 \\ 1 \end{bmatrix} + c_2 \begin{bmatrix} 0 \\ e^{-kt} \\ -e^{-kt} \end{bmatrix} + c_3 \begin{bmatrix} e^{-kt} \\ (kt-1)e^{-kt} \\ -kte^{-kt} \end{bmatrix}.$$

Using the boundary conditions yields

$$c_1 = 2\left(1 - e^{-k(T_1 + T_2 + T_3)}\right);$$
$$c_2 = 2\left(1 - e^{-k(T_1 + T_2)}\right);$$

and
$$c_3 = 2(1 - e^{-kT_1}).$$

Therefore, the final form of the solution can be written as

$$I(t) = \begin{bmatrix} I_1 \\ I_2 \\ I_3 \end{bmatrix} = 2(1 - e^{-k(T_1 + T_2 + T_3)}) \begin{bmatrix} 0 \\ 0 \\ 1 \end{bmatrix}$$

$$+ 2(1 - e^{-k(T_1 + T_2)}) \begin{bmatrix} 0 \\ e^{-kt} \\ -e^{-kt} \end{bmatrix}$$

$$2(1 - e^{-kT_1}) \begin{bmatrix} e^{-kt} \\ (kt - 1)e^{-kt} \\ -kte^{-kt} \end{bmatrix}.$$

9. Sample Exam

1. Given that the initial concentrations of a red dye are 0.04 and 0.10 for compartments 1 and 2 respectively, which have volumes of 10 and 20 liters, and that the coefficient of permeability between the two compartments is $k_{12} = k_{21} = 0.25$, determine:
   a. the concentration $c_i(t)$ in each compartment at time t;
   b. the concentration of dye in compartment 1 when equilibrium occurs.

2. Use the data in problem 1 above, except that now assume $k_{12} = 0.25$ but $k_{21} = 0.05$.
   a. Find the concentration $c_i(t)$ in each compartment.
   b. Determine $c_1(\infty)$ and $c_2(\infty)$ and explain why they are not equal.

3. Use the generalized compartmental model discussion in section 5.2 to obtain the system of differential equations corresponding to the following problem. A compartmental system consists of three compartments. Two chemicals α_1 and α_2 are present in the system. The compartments are assumed to have constant volumes and be well stirred; also, we assume $\rho_{kj} = \rho_j$. We further assume no interaction of the α_i with the compartments and that no diffusion occurs. Find the system of equations describing the situation if $Q_{ij} = j - i$, $\rho_k^{\alpha_1} = kt$, and $\rho_k^{\alpha_2} = 2kt$. Note you will need to use equation (19),

$$\dot{y}_k^\alpha = \sum_i \rho_i^\alpha Q_{ik} - \sum_j \rho_j^\alpha Q_{kj} + \sum_h D_{nk}^\alpha + v_k r_k^\alpha,$$

for each species α.

4. For each of the following compartmental matrices, determine all states j from which the process can exit the system. In each case, also calculate the corresponding exit coefficients k_{0j}.

a. $\begin{bmatrix} -1 & 1 & 0 & 0 \\ 1 & -2 & 1 & 1 \\ 0 & 0 & -1 & 0 \\ 0 & 0 & 0 & -1 \end{bmatrix}$ b. $\begin{bmatrix} -3 & 2 & 0 \\ 1 & -2 & 1 \\ 1 & 0 & -2 \end{bmatrix}$

10. Answers to Sample Exam

1. a.

$$c_1(t) = 0.08 - 0.04 \exp[-0.0375t]$$
$$c_2(t) = 0.08 + 0.02 \exp[-0.0375t]$$

b. $c_1(\infty) = 0.08$

2. a.

$$c_1(t) = 0.0218 + 0.0182 \exp[-0.0275t]$$
$$c_2(t) = 0.1091 - 0.0091 \exp[-0.0275t]$$

b.

$$c_1(\infty) = 0.0218$$
$$c_2(\infty) = 0.1091$$

The equilibrium concentrations are not equal because the coefficients of permeability are not equal and the volumes are not equal for the compartments.

3. We use (21) to get

$$\dot{y}_k^{\alpha_1} = \sum_i \rho_i^{\alpha_1} Q_{ik} - \sum_j \rho_k^{\alpha_1} Q_{kj}$$
$$= t(k-1) + 2t(k-2) + 3t(k-3)$$
$$- kt(1-k) - kt(2-k) - kt(3-k)$$
$$\boxed{\dot{y}_k^{\alpha_1} = -14t + 3k^2 t}\ ;$$

therefore

$$\boxed{\dot{y}_1^{\alpha_1} = -14t + 3t = -11t}$$
$$\boxed{\dot{y}_2^{\alpha_1} = -14t + 3(4)t = -2t} \quad (25)$$
$$\boxed{\dot{y}_3^{\alpha_1} = -14t + 3(3)^2 t = 13t}$$

Similarly, for species α_2 we get

$$\dot{y}_k^{\alpha_2} = \sum_i \rho_i^{\alpha_2} Q_{ik} - \sum_j \rho_k^{\alpha_2} Q_{kj}$$

$$= 2t(k-1) + 4t(k-2) + 6t(k-3)$$
$$- 2kt(1-k) - 2kt(2-k) - 2kt(3-k)$$

$$\boxed{\dot{y}_k^{\alpha_2} = -28t + 6k^2 t}\ ;$$

therefore

$$\boxed{\begin{aligned}\dot{y}_1^{\alpha_2} &= -28t + 6t = -22t \\ \dot{y}_2^{\alpha_2} &= -28t + 6(2)^2 t = -4t \\ \dot{y}_3^{\alpha_2} &= -28t + 6(3)^2 t = 26t\end{aligned}}$$

(26)

The six equations given by (25) and (26) describe the compartmental system for this problem.

4. a. Only the second column does not sum to 0. Hence

$$k_{22} = -2 = -\sum_{i \neq 2} k_{i2} - k_{02} = -1 - k_{02}.$$

Solving gives $k_{02} = 1$; i.e., the only exit is via state 2 with exit coefficient of 1.

b. Columns 1 and 3 both sum to negative numbers; hence exit is possible from either state 1 or state 3.

$$k_{11} = -3 = -\sum_{i \neq 1} k_{i1} - k_{01} = -2 - k_{01} \Rightarrow \boxed{k_{01} = 1}.$$

Similarly,

$$k_{33} = -2 = -\sum_{i \neq 3} k_{i3} - k_{03} = -1 - k_{03} \Rightarrow \boxed{k_{03} = 1}.$$

11. References

Anderson, David H. 1983. *Compartmental Models and Tracer Kinetics*. New York: Springer-Verlag.

⎯⎯⎯. 1979. Interactive inversion of single exit compartmental matrices. *Computational Biological Medicine* 9:317–330.

⎯⎯⎯, et al. 1977. The mathematical analysis of a four compartment stochastic model of Rose Bengal transport through the hepatic system. In *Nonlinear Systems and Applications*. New York: Academic Press.

Atkins, G.L. 1969. *Multicompartment Models for Biological Systems*. London: Methuen.

Banks, H.T. 1979. *Lectures on Differential Equation Models in Biology*. Lefschetz Center for Dynamical Systems, Brown University.

_____ and D. Palatt. 1975. *Mathematical Modeling in the Biological Sciences*. LCDS Lecture Notes 75-1. Lefschetz Center for Dynamical Systems, Brown University.

Batschelet, E. et al. 1979. On the kinetics of lead in the human body. *Journal of Mathematical Biology* 8:15–23.

An excellent and readable use of modeling techniques is developed in this module.

Cullen, M.R. 1985. *Linear Models in Biology: Linear Systems Analysis with Biological Applications*. New York: John Wiley and Sons.

This is the best place for readers finishing this module to continue.

Dosik, G.M., et al. 1986. A rapid automated stathmokinetic method for determination of in vitro cell cycle transit times. *Cell Tissue Kinetics* 14:121–134.

Farnsworth, C.H. 1978. Comparing numerical methods for solving linear systems of equations involving single exit compartmental systems. Unpublished report, Southern Methodist University.

Finizio, N. and G. Ladas. 1978. *Ordinary Differential Equations with Modern Applications*. Belmont, CA: Wadsworth Press.

Godfrey, Keith. 1983. *Compartmental Models and Their Application*. New York: Academic.

Jacquez, J.A. 1972. *Compartmental Analysis in Biology and Medicine*, Holland: Elsevier.

Randall, James E. 1980. *Microcomputers and Physiological Simulation*. Reading, MA: Addison-Wesley.

Professor Randall has programmed many instructional simulations, featuring compartment models, which he distributes at cost. These include "Physiological Simulations," "BASIC Electro-physiology," and "BASIC Cardiovascular," with student worksheets. Interested readers may contact him at the Indiana University School of Medicine, Bloomington, IN 47405.

Spain, James D. 1982. *Basic Microcomputer Models in Biology*. Reading, MA: Addison-Wesley.

A disk of programs in Apple II format is available with the text.

About the Author

Ron Barnes received his undergraduate degree in mathematics from St. Bonaventure University and master's and Ph.D. degrees from Syracuse University. He is associate professor of Applied Mathematical Sciences at the University of Houston, Downtown. He is a visiting lecturer for the MAA and past president of the Houston chapter of the American Statistical Association. His main areas of interest are statistics, applied mathematics with applications in biology, and OR/decision theory.

UMAP Module 677

Modules in Undergraduate Mathematics and its Applications

Funding Pension Benefits: The Individual Spread Gain Method

Ho Kuen Ng

Published in cooperation with the Society for Industrial and Applied Mathematics, the Mathematical Association of America, the National Council of Teachers of Mathematics, the American Mathematical Association of Two-Year Colleges, The Institute of Management Sciences, and the American Statistical Association.

COMAP

INTERMODULAR DESCRIPTION SHEET:	UMAP Unit 677
TITLE:	FUNDING PENSION BENEFITS: THE INDIVIDUAL SPREAD GAIN METHOD
AUTHOR:	Ho Kuen Ng Department of Mathematics and Computer Science San Jose State University San Jose, CA 95192 with a practitioner's commentary by Thomas R. Eckert, F.S.A. Associate Actuary, Employee Benefits Division CUNA Mutual Insurance Group 5910 Mineral Point Road P.O. Box 391 Madison, WI 53701-0391
MATHEMATICAL FIELD:	Precalculus
APPLICATION FIELD:	Finance, insurance, actuarial science
TARGET AUDIENCE:	Students in courses in business, finance, precalculus, or mathematics of finance
ABSTRACT:	The individual spread gain method (also known as the individual aggregate method) of funding pension benefits is described, with examples and a practitioner's commentary.
PREREQUISITES:	High school algebra, plus concept of present value

© Copyright 1987 by COMAP, Inc. All rights reserved.

COMAP, 60 Lowell Street, Arlington, MA 02174 (617) 641-2600

Funding Pension Benefits: The Individual Spread Gain Method

Ho Kuen Ng
Department of Mathematics and Computer Science
San Jose State University
San Jose, CA 95192

Table of Contents

1. INTRODUCTION 1
2. BASIC EXAMPLE 2
3. SALARY-DEPENDENT BENEFIT 3
4. AMORTIZATION AS A PERCENTAGE OF SALARY 4
5. A LARGER PENSION PLAN 6
6. REFERENCES 8
7. ANSWERS TO EXERCISES 8
APPENDIX: PRACTITIONER'S COMMENTARY
     by Thomas R. Eckert 9

Modules and Monographs in Undergraduate Mathematics and Its Applications (UMAP) Project

The goal of UMAP was to develop, through a community of users and developers, a system of instructional modules in undergraduate mathematics and its applications to be used to supplement existing courses and from which complete courses may eventually be built.

The Project was guided by a National Advisory Board of mathematicians, scientists, and educators. UMAP was funded by a grant from the National Science Foundation and is now supported by the Consortium for Mathematics and Its Applications (COMAP), Inc. a nonprofit corporation engaged in research and development in mathematics education.

COMAP Staff

Solomon A. Garfunkel	Executive Director, COMAP
Laurie W. Aragon	Business Development Manager
Philip A. McGaw	Production Manager
Nancy Hawley	Copy Editor
Annemarie S. Morgan	Administrative Assistant

1. Introduction

The funding for a pension plan means the deposit of contributions to the plan periodically to pay for the retirement benefits of plan participants. The sole purpose of funding for a pension plan is to make sure that money is accumulated in a systematic way so that a plan participant gets what the participant is entitled to by the time of retirement.

The attraction of a pension plan to an employee is obvious. On the other hand, why does an employer want to establish a pension plan? To attract and retain talents, to retire career employees in a dignified and systematic way, to satisfy union demands, to take tax advantages: these are among the more important reasons.

How is a pension plan funded? Usually an actuary is responsible for making the necessary calculations. An actuary is a person trained in the mathematics of pension and insurance systems. There are several funding methods that are allowed by U.S. law. In this module, we will concentrate on the Individual Spread Gain actuarial cost method, also known as the Individual Aggregate method. The fancy name will become more comprehensible after we understand how the method works.

Notation

Let i denote the interest rate.

$v = 1/(1 + i)$, the discount factor.

$\ddot{a}_{\overline{n}|} = 1 + v + v^2 + v^3 + \cdots + v^{n-1}$, the present value at the beginning of year 1 of a stream of cash flow of \$1 for n years, deposited at the beginnings of years 1 through n.

$\ddot{s}_{\overline{n}|} = (1 + i)^n + (1 + i)^{n-1} + \cdots + (1 + i)$, the future value at the end of year n of a stream of cash flow of \$1 for n years, deposited at the beginnings of years 1 through n.

Note that these are standard notations in actuarial mathematics. The dots do not refer to derivatives, as are sometimes used in calculus.

The ideas presented in this article are a very simplified version of the method used for funding many small pension plans, especially for professionals. Readers interested in this or other actuarial cost methods, or actuarial mathematics in general, may consult the references listed at the end of the module.

In all the numerical examples and exercises involving dollar amounts, the amounts are rounded to the nearest dollar.

2. Basic Example

To make things easy, let us start with a very simple example. Assume that there is only one participant in the pension plan, who is 35 years of age on 1/1/87. The plan provides that the participant may retire at age 65 and receive a monthly benefit of $1000. Suppose that a monthly annuity of $1 from age 65 for life costs $140. Here the cost of a monthly annuity of $1 from age x for life is an amount, on the average, that is sufficient, together with interest, to pay a monthly benefit of $1 to a person x years of age, until death. Then the amount of money that will be needed when the plan participant retires at age 65 is $1000 \times 140 = \$140{,}000$. If we plan to set aside a constant amount of money at the beginning of each year, how much is the dollar amount needed? Let $\ddot{s}_{\overline{n}|}$ denote the future value at the end of year n of a stream of cash flow of $1 for n years, deposited at the beginnings of years 1 through n. If A is set aside at the beginning of each year, the fund will accumulate to $\$A \times \ddot{s}_{\overline{30}|}$ at the participant's retirement.

Assuming an interest rate of 6%, and equating the two quantities, we have $A = 140{,}000/\ddot{s}_{\overline{30}|} = 140{,}000/83.8017 = 1671$. This amount is called the *normal cost* for the year.

Instead of future values, we may work with present values (and this has been done traditionally). Let $\ddot{a}_{\overline{n}|}$ denote the present value at the beginning of year 1 of a stream of cash flow of $1 for n years, deposited at the beginnings of years 1 through n. The present value of the needed fund is $\$140{,}000/1.06^{30} = \$24{,}375$. We then have $A \times \ddot{a}_{\overline{30}|} = 24{,}375$, also giving $A = 1671$.

If everything works out as expected, then an annual contribution of $1671 will give the required $140,000 after 30 years. Let us check it after one year, i.e., as of 1/1/88, when the plan participant is 36 years old. The present value of the retirement benefit then is $\$140{,}000/1.06^{29} = \$25{,}838$. The contribution put in at the beginning of the previous year accumulates, with interest, to $1771. Thus we need to accumulate, over 29 years, an amount with present value $\$25{,}838 - \$1771 = \$24{,}067$. Solving the equation $A \times \ddot{a}_{\overline{29}|} = 24{,}067$, we have $A = 24{,}067/14.4062 = 1671$, as expected.

What if the money gains more than 6%? Suppose the assets of the pension fund earn 10% in the year 1987. Then the fund as of 1/1/88 amounts to $1838, and so we need only accumulate a present value of $24,000 over 29 years. The solution of a similar equation gives the normal cost $24{,}000/\ddot{a}_{\overline{29}|} = \$1666$.

Can we explain the difference of $5 in the normal costs? Yes, and very easily. The assets in the pension fund are supposed to earn 6%, but in fact they earn 10%. This difference is called the investment gain, which is $1671 \times (10\% - 6\%) = \$67$. Note that this is the difference between the numerators 24,067 and 24,000. Consequently the difference of $5 in the normal costs is $67/\ddot{a}_{\overline{29}|}$. In other words, we are *spreading the gain* over the remaining working lifetime of the employee. Hence the name Spread Gain actuarial cost method.

What if we have an investment loss, i.e., earnings less than 6%? We spread the loss too. (See **exercise 1**.)

An advantage of this method is that it is self-correcting. If we make a mistake in one year, the error will appear as a gain or loss in the next year, and will then be spread over the remaining working lifetime of the plan participant. Its effect will not be as painful as to have to recognize it all at once in the normal cost. On the other hand, we are still sure that a sufficient fund will be ready at the employee's retirement.

"*...we are spreading the gain over the remaining working lifetime of the employee.*"

Exercises

1. In the basic example, if the pension fund as of 1/1/88 amounts to $1500, find the investment loss and normal cost for the year 1988.

2. (Continuation of **Exercise 1**.) If the assumptions are exactly realized in the year 1988, show that the normal cost for the year 1989 is the same as that of the year 1988.

3. Salary-Dependent Benefit

With a high inflation rate, a flat amount of benefit, say $1000 per month, may not mean very much when the employee retires. Also, more highly paid employees expect to get higher retirement benefits. To accommodate these facts, many plans give a retirement benefit of some percentage of the employee's salary during the final year of employment. Presumably, salary should keep up with inflation; so retirement benefit based on the final year's salary should reflect the cost of living at the time of the employee's retirement. Given that this is a good approach, how can the pension actuary determine the normal cost, since at the time of calculation, there can be no certainty what the employee's final salary will be? What is usually done is to assume a salary scale, based on the salary history of the employee, the personnel policy of the company, and the actuary's forecast of future economic conditions.

Let us assume that a salary increment rate of 7% per year is used, and that the monthly retirement benefit is 50% of the monthly salary in the final year of employment. Continuing with the basic example, we suppose that the employee earns $24,000 in the year 1987. Then the final year's salary, according to assumption, will be $24,000 \times 1.07^{29} = \$170,742$. Note that the exponent is 29, not 30

($= 65 - 35$), because the final year of employment is the year beginning when the employee just reaches 64 years of age.

Now the retirement benefit per month = $\$170{,}742 \times 50\%/12 = \$7114$, with a present value of $\$7114 \times 140/1.06^{30} = \$173{,}413$. This gives a normal cost of $\$173{,}413/\ddot{a}_{\overline{30|}} = \$11{,}885$.

After one year, if the assumptions are exactly realized, the employee's salary for the year 1988 is $\$24{,}000 \times 1.07 = \$25{,}680$, with a projected monthly benefit of $\$25{,}680 \times 1.07^{28} \times 50\%/12 = \$7114$. This has a present value of $\$7114 \times 140/1.06^{29} = \$183{,}818$. As of 1/1/88, the pension fund should have accumulated to $\$11{,}885 \times 1.06 = \$12{,}598$. This gives a normal cost equal to ($\$183{,}818 - \$12{,}598)/\ddot{a}_{\overline{29|}} = \$11{,}885$, as expected.

The previous method is sometimes called amortization using a constant (or level) dollar amount, the reason being that the normal cost each year, if the assumptions are exactly realized, is constant.

Exercises

(Refer to the example in this section.)

3. If the assets in the pension fund earn 10% instead of 6% in the year 1987, and salary increases only 5%, calculate the normal cost for the year 1988.

4. If the employer decides to liberalize the retirement benefit to 52% as of 1/1/88, and that the actuarial assumptions are exactly realized in 1987, find the normal cost for the year 1988.

($\$183{,}818 - \$5330$)/(present value of salaries from 1987) $\times$ 1987 salary,

4. Amortization as a Percentage of Salary

Another approach to fund for a pension plan is to amortize as a constant percentage of salary. Why does the salary rise? Part of the reason may be inflation. Due to inflation, a constant annual contribution means a much heavier burden to the employer in the present than in the future. Another part of the reason may be the financial strength of the company. The salaries paid may be a rough guideline of the strength of the employer to contribute. In other words, the higher the salary of an employee, the more we expect the employer will be willing and able to contribute to the pension plan on behalf of the employee.

Using our previous example, the present value of retirement benefit as of 1/1/87 is $173,413. What is the present value of the future salary stream? Applying the salary increment rate and the discount factor, we have

the present value of future salaries = $24,000 + $24,000 × 1.07/1.06 + $24,000 × $1.07^2/1.06^2$ + ⋯ + $24,000 × $1.07^{29}/1.06^{29}$ = $24,000 × $((1.07/1.06)^{30} - 1)$/$((1.07/1.06) - 1)$ = $827,743.

The ratio of the present value of retirement benefit to the present value of future salaries is 0.2095, or 20.95%.

We claim that if the employer contributes this percentage of the employee's salary at the beginning of each year, and if the assumptions are exactly realized, then the accumulated contributions will just be able to pay the retirement benefit. To prove this, assume that the required percentage is P, and the salary increment and interest rates are r and i, respectively. Also let the current salary be $$S$. Then we have the equation

$$P \times S + P \times S \times (1+r)/(1+i)$$
$$+ P \times S \times (1+r)^2/(1+i)^2 + \cdots$$
$$= \text{present value of retirement benefit.}$$

This simplifies to

P × present value of future salaries
= present value of retirement benefit,

and so

P = present value of retirement benefit/present value of future salaries,

as claimed.

Now let us apply this result to our numerical example. The first year's normal cost = $24,000 × 20.95% = $5028.

If the assumptions are exactly realized, the fund should accumulate to $5330 as of 1/1/88. By that time, the present value of retirement benefit = $183,818, salary of the year 1988 = $25,680, and the normal cost will then be

($183,818 − $5330)/(present value of salaries from 1988) × 1988 salary,

according to the theory. This works out to be $5380.

Let us observe that this is precisely 20.95% of the salary of the year 1988 of $25,680, confirming what we have proved.

Here, we can also appreciate the long-term commitment of an employer in establishing a pension plan for employees. By promising a 50% continuation of salary during retirement years, the employer is essentially giving a salary increase of 20.95%.

What if the assets in the pension fund amount to $10,000 as of 1/1/88, i.e., with an investment gain of $4670? In this case, we need only accumulate $183,818 − $10,000 = $173,818 in the remaining 29 years. The present value of future salaries is

$$\$25{,}680 + \$25{,}680 \times 1.07/1.06 + \$25{,}680 \times 1.07^2/1.06^2 + \cdots + \$25{,}680 \times 1.07^{28}/1.06^{28}$$
$$= \$851{,}968.$$

This gives a normal cost percentage of 20.4% and a normal cost of $5239. We see that the gain of the fund lowers the normal cost percentage from 20.95% to 20.4%, and the normal cost from $5380 to $5239. By lowering the normal cost percentage, the investment gain is spread over the remaining working lifetime of the employee.

Note that if the retirement benefit is not salary-related, U.S. law does not allow amortization as a percentage of salary. If the Individual Spread Gain actuarial cost method is to be used for such a pension plan, the actuary *must* use amortization as a level dollar amount.

Exercises

5. As of 1/1/87, an employee is 40 years of age and earns an annual salary of $36,000. The pension plan provides for a retirement age of 60 and a benefit of 25% of the final year's salary. Assuming the interest rate of 6%, a salary increment rate of 6.5%, and the cost of a monthly annuity of $1 from age 60 for life of $160, calculate the normal cost for the year 1987.

6. (Continuation of **exercise 5.**) In the year 1987, the fund earns 5% and salary increases by 10%. Calculate the normal cost percentage for the year 1988. Is this lower or higher than the normal cost percentage for the year 1987? Can you explain why?

7. Show that, if the interest rate and the salary increment rate are equal, then the result is exactly the same as if both the interest and salary increment rates are 0%, at the end of the first year.

5. A Larger Pension Plan

Of course, most pension plans cover more than one single employee. In this section, we will generalize our funding method to a larger plan.

Again we illustrate the idea by using an example. There are two employees participating in the pension plan. As of 1/1/87, X is 35 years old and earns an annual salary of $24,000, while Y is 45 years of age with an annual salary of $30,000. The plan provides for a retirement age of 65 and a monthly retirement benefit of 30% of the monthly salary in the last year of employment. The actuary uses the level dollar amortization method, a 6% interest rate, and a 7% salary increment rate.

The reader can easily verify the following numbers.

	X	Y
Final year's salary	$170,742	$108,496
Projected monthly benefit	4269	2712
Amount needed at retirement	597,598	379,735
Present value of above amount	104,048	118,403
n = retirement age − current age	30	20
$\ddot{a}_{\overline{n}\rceil}$	14.5907	12.1581
Normal cost	7131	9739

If all actuarial assumptions are exactly realized, the pension fund as of 1/1/88, with interest, should be $17,882. X and Y should have salaries of $25,680 and $32,100 for the year 1988. It is an easy exercise to show that the total normal cost for the year 1988 is then $16,870, as in the year 1987.

In reality, we cannot expect all the assumptions to be realized precisely. Let us assume that the assets of the pension fund as of 1/1/88 have a value of $20,000 (a gain), X's salary for the year 1988 is $25,000 (a gain for the plan, because the salary, and so the benefit, are not as high as projected in 1987), and Y's salary is $35,000 (a loss for the plan).

The first step in the 1/1/88 calculation is to allocate the fund of $20,000 between X and Y. According to the previous year's calculation, X and Y should have $7131 and $9739 under their accounts as of 1/1/87. Now we simply allocate the $20,000 in that proportion. In other words, we calculate what the actual interest rate is, and apply the rate to the money in the individual participant's account. As a result, X and Y are allocated $8454 and $11,546, respectively as of 1/1/88.

A straightforward calculation yields the following.

	X	Y
Final year's salary	$166,221	$118,298
Projected monthly benefit	4156	2957
Amount needed at retirement	581,773	414,042
Present value of above amount	107,370	136,846
Current allocated fund	8454	11,546
Amount to be made up	98,916	125,300
n = retirement age − current age	29	19
$\ddot{a}_{\overline{n}\rceil}$	14.4062	11.8276
Normal cost	6866	10,594

By dividing by $\ddot{a}_{\overline{n}|}$, we are essentially spreading the gains and losses due to deviations from the actuarial assumptions over the remaining working lifetimes of the individual participants. Hence the name Individual Spread Gain actuarial cost method.

Exercises

8. Redo the example in this section as of 1/1/87, using amortization as a percentage of salary.

9. With the deviations from the actuarial assumptions in the year 1987 as in the example in this section, calculate the total normal cost for the year 1988, using amortization as a percentage of salary.

6. References

Anderson, A.W. 1984. *Pension Mathematics*. Windsor Press.

Bowers, N.L., et al. *Actuarial Mathematics*. Society of Actuaries.

Berin, B.N. 1978. *The Fundamentals of Pension Mathematics*. William M. Mercer, Inc.

Kellison, S.G. 1970. *The Theory of Interest*. Homewood, IL: Richard D. Irwin, Inc.

McGill, D.M. 1984. *Fundamentals of Private Pensions*, 5th ed. Homewood, IL: Richard D. Irwin, Inc.

Shapiro, Arnold F. 1983. Modified cost methods for small pension plans. *Transactions of the Society of Actuaries* 35.

Trowbridge, C.L., and C.E. Farr. 1976. *The Theory and Practice of Pension Funding*. Homewood, IL: Richard D. Irwin, Inc.

Winklevoss, H.E. 1977. *Pension Mathematics with Numerical Illustrations*. Homewood, IL: Richard D. Irwin, Inc.

About the Author

Ho Kuen Ng received his B.S. degree from the University of Hong Kong and his Ph.D. degree from the University of California at Berkeley. Besides teaching at San Jose State University, he also spends some of his time doing actuarial work. His major interests are algebra, operations research, and actuarial science.

7. Answers to Exercises

1. $338
   $1689
2. $1689

3. $11,614
4. $12,396
5. $5917
6. $6522
   Fund earns less, salary increases more—both are losses to the plan.
7. easy algebra
8. $3017 for X, $5407 for Y
9. $3020 for X, $5990 for Y

Appendix: Practitioner's Commentary

Thomas R. Eckert
Associate Actuary, Employee Benefits Division
CUNA Mutual Insurance Group
5910 Mineral Point Road
P.O. Box 391
Madison, WI 53701-0391

Professor Ng offers an accurate description of the Individual Spread Gain method. Although not a popular one, this method is acceptable for determining the accounting and tax figures in funding a pension plan. It is easily explained to a plan sponsor, though other methods offer simpler calculation. An analysis of some of the advantages and disadvantages of the Individual Spread Gain method follows.

The Individual Spread Gain cost method is a *Cost Allocation cost method* without supplemental liability. This means that the plan projects plan benefits and then allocates the present value of those benefits over the future working lifetimes of the plan participants. Other cost-allocation methods spread out experience gains and losses, or past service liabilities arising from the adoption of a plan, over specific amortization periods. The method by which these supplemental liabilities are calculated also varies by the cost method chosen.

There is another family of methods, *Benefit Allocation cost methods*, which calculates a specific benefit associated with each year of service and allocates the cost of that benefit accordingly. These methods tend to generate costs that increase with age due to the higher salaries at later ages, the older age of the participant, and the reduced time over which to fund the benefit. The cost-allocation methods, on the other hand, generally produce recommended costs that are a stable percentage of compensation; this is usually a desirable result for the budgeting of plan costs [McGill 1984].

There is a wide variety of cost methods that are utilized in U.S. practice. The criteria [Shapiro 1983] that are used to choose a method may consist of:

1. Adequacy of the fund compared to the present value of the accrued benefits at any point in time.

2. Consistency of the cost from year to year.

3. Flexibility of the plan in determining recommended deposits.

4. Robustness of the plan and its ability to handle gains and losses from experience.

5. Acceptability of the method to the U.S. Internal Revenue Service (IRS).

6. Simplicity of the method.

Of these criteria, the Individual Spread Gain method provides no guarantee of the first item, since the costs of the plan are not directly related to the accrued benefit. Each year's contributions are based on the projected benefits at retirement and are not compared to the benefit that the participant currently has accrued.

The second criterion—consistency of cost—is not provided for by the method. When employed with realistic assumptions, the method will provide a level contribution from year to year, either as a percentage of payroll or as a dollar amount, for young participants only. For older participants, however, the method produces costs that can fluctuate.

The method is not flexible in allowing the plan sponsor to have a range of contributions from which to fund the plan. Rather, it determines only a single recommended contribution. Other methods produce a separate calculation for past service costs, amendments, changes in actuarial assumptions, and gains and losses, and allow these costs to be amortized over a variable number of years, which provides flexibility in funding.

An advantage of the Individual Spread Gain method is that gains and losses, due to plan experience that differs from the assumptions used by the plan actuary, are amortized directly over the future working lifetime of the plan participant. For older participants, this may produce large swings in the calculation of the pension contribution, which would be an undesirable result; for younger participants this would not be a problem.

The method is automatically acceptable by the IRS for use in determining minimum funding amounts and maximum allowable contributions.

The Individual Spread Gain method can be easily explained to plan sponsors. It is generally used for small plans of one to two participants, as the cost for a larger group can be more efficiently calculated by using a group cost method. The method is also more difficult to apply because asset accounting has to be collected for each individual participant. Utilizing a Group Aggregate method would allow the plan actuary to employ the entire fund balance for the group and avoid this accounting.

Another advantage of the Individual Spread Gain method is that there is an explicit cost for each individual participant that may be communicated to the plan sponsor. Some plan sponsors are accustomed to knowing the exact cost of benefits for each participant, and some group methods do not directly provide this accounting.

At CUNA Mutual we do not currently apply the Individual Spread Gain method to fund any of the 1200 small Defined Benefit pension plans for which we provide actuarial services. Instead, the majority of our plans are funded through the Aggregate and Modified Aggregate methods. We also use the Frozen Initial Liability method for plans that want to pay for past service that has been accumulated prior to the adoption of the plan. This method is similar to the Spread Gain method, with the addition that it identifies a separate cost for past service and allows it to be amortized independently of the cost of each current year's accrual of benefits. (These methods are all explained in [McGill 1984].)

About the Author

Ho Kuen Ng received his B.S. degree from the University of Hong Kong, and his Ph.D. degree from the University of California at Berkeley. Besides teaching at San Jose State University, he also spends some of his time doing actuarial work. His major interests are algebra, operations research, and actuarial science.

UMAP
Module 679

Modules in
Undergraduate
Mathematics
and Its
Applications

The Solar Concentrating Properties of a Conical Reflector

Don Leake

Published in
cooperation with
the Society
for Industrial
and Applied
Mathematics, the
Mathematical
Association of
America, the
National Council
of Teachers of
Mathematics,
the American
Mathematical
Association of Two-
Year Colleges, and
The Institute
of Management
Sciences.

COMAP

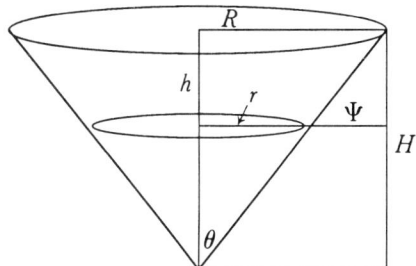

INTERMODULAR DESCRIPTION SHEET:	UMAP Unit 679
TITLE:	THE SOLAR CONCENTRATING PROPERTIES OF A CONICAL REFLECTOR
AUTHOR:	Don Leake Department of Mathematics/Computer Science University of Wisconsin, River Falls River Falls, WI 54022
APPLICATION FIELD:	Solar energy optics
ABSTRACT:	A series of 33 problems guides the reader in an analysis of the effect of adding a conical reflecting surface to a circular solar radiation collecting disk. Considered first is the problem of characterizing the amount of solar energy absorbed by the conical collector when it is aimed directly at the sun. This simpler problem leads to an investigation of the performance of the solar collector when it is not aimed directly at the sun. Finding the solution of the first problem incorporates elements of algebra, trigonometry, and calculus. The more general problem requires knowledge of analytic geometry and elementary vector analysis. An analytic solution to the first problem is found, while the second requires the development of a computer program.
PREREQUISITES:	Algebra, trigonometry, basic calculus, analytic geometry, elementary vector analysis, and computer programming.

© Copyright 1982, 1987 by COMAP, Inc. All rights reserved.

COMAP, 60 Lowell Street, Arlington, MA 02174 (617) 641-2600

The Solar Concentrating Properties of a Conical Reflector

Don Leake
Department of Mathematics/Computer Systems
University of Wisconsin, River Falls
River Falls, WI 54022

Table of Contents

1. INTRODUCTION 1
   1.1 The Use of a Conical Reflector
       in a Solar-Powered Engine Design 1
   1.2 Statement of Objectives 2
2. FUNDAMENTAL CONCEPTS 3
   2.1 Characterization of the Reflecting Cone 3
   2.2 Characterization of the Energy Carried
       by Solar Rays 4
   2.3 Power Passing through the Aperture of the Cone 5
3. DIRECT RAY ANALYSIS 6
   3.1 Fundamental Law of Reflection 6
   3.2 Cone Parameters 7
   3.3 Radii of Reflection 8
   3.4 Concentrating Properties of Ideal Conical Reflectors ... 10
   3.5 Nonideal Reflectors 11
4. OBLIQUE RAY ANALYSIS 12
   4.1 Incoming Ray Description 12
   4.2 Partitioning of the Aperture 13
   4.3 Computing the Power of the Collector 14
   4.4 Analytic Description of the Cone 14
   4.5 Intersection of an Incoming Ray with the Cone 15
   4.6 Intersection with the Collecting Disk 17
   4.7 Inward Normal to the Cone Surface 17
   4.8 Computation of the Direction of the Reflected Ray 17
   4.9 New Intersection Point 17
   4.10 Power Magnification 18
5. SUMMARY 20
6. ANSWERS TO THE PROBLEMS 20
   REFERENCES

MODULES AND MONOGRAPHS IN UNDERGRADUATE
MATHEMATICS AND ITS APPLICATIONS (UMAP) PROJECT

The goal of UMAP was to develop, through a community of users and developers, a system of instructional modules in undergraduate mathematics and its applications to be used to supplement existing courses and from which complete courses may eventually be built.

The Project was guided by a National Advisory Board of mathematicians, scientists, and educators. UMAP was funded by a grant from the National Science Foundation and is now supported by the Consortium for Mathematics and Its Applications (COMAP), Inc., a nonprofit corporation engaged in research and development in mathematics education.

COMAP Staff

Paul J. Campbell	Editor
Solomon A. Garfunkel	Executive Director, COMAP
Laurie W. Aragon	Business Development Manager
Philip A. McGaw	Production Manager
Theresa Cronin	Copy Editor
Annemarie S. Morgan	Administrative Assistant
John Gately	Distribution

Acknowledgment

The author is grateful to the Lilly Endowment for a fellowship which made this work possible. Also, the author is indebted to J.R. Senft for his inspiration and guidance throughout this project.

1. Introduction

1.1 The Use of a Conical Reflector in a Solar-Powered Engine Design

In 1985 J.R. Senft suggested the possibility of powering Stirling engines with moderately concentrated solar energy. Unlike the internal combustion engines found in most automobiles, Stirling engines operate on a closed cycle and are externally heated. Senft developed an experimental Ringbom Stirling engine powered solely by a circular solar collecting disk [Senft 1986]. A conical reflector was incorporated in his design, to enhance the engine's solar collecting capabilities.

The use of a conical reflector in a solar-collecting design dates back to one of the earliest solar engines, developed by the pioneering Frenchman August Mouchot in 1875 [Meinel and Meinel 1976]. Mouchot's device was a low-pressure steam engine that employed a conical reflecting surface to focus the sun's rays uniformly on an energy-absorbing cylinder at the cone's axis. The thermal efficiency of Mouchot's engine, the fraction of thermal energy converted to mechanical energy by the engine, was lower than the thermal efficiencies of conventional coal-fueled steam engines, mainly because the engine operated at a lower temperature range.

In its regenerative form, a Stirling engine's thermal efficiency matches the Carnot thermal efficiency [Lee and Sears 1963]. The Carnot thermal efficiency is an upper bound to all thermal efficiencies of closed-cycle engines and is given by the formula $E = (T_H - T_C)/T_H$, where T_H and T_C represent respectively the hottest and coldest temperatures attained during the cycle. Attempts at increasing the thermal efficiency of closed-cycle engines are often guided by the Carnot thermal efficiency formula. The quantity T_C is usually fixed at the ambient temperature; in such a case, E can only be improved by increasing T_H. Senft's engine achieved a higher T_H with the addition of a conical reflector.

Solar furnaces and solar ovens can achieve higher temperatures by the use of parabolic or spherical reflectors to focus the sun's radiation at one small spot. Higher temperatures require the use of more expensive heat-resistant construction materials. In addition, these focusing solar reflectors are very sensitive to the position of the sun and must be furnished with elaborate tracking mechanisms to follow the sun's path. A truncated conical reflector, which directs solar rays to a collecting disk at its base, provides moderate temperature increases and is relatively insensitive to the sun's position; it can be left standing for long periods of time without adjustment. Because Stirling engines are able to operate at low temperature differentials, the easily built conical reflector is a logical selection as a solar concentrator for this type of engine.

1.2 Statement of Objectives

This study investigates the effect of adding a conical reflecting surface to a circular solar collecting disk [Burkhard et al. 1977]. The analysis is accomplished by the reader, who is guided by a series of 33 problems. The answers to these problems are given in section 6. The problem of computing the power of the solar radiation absorbed by the collecting disk is solved by determining which of the incoming rays are reflected to the absorption disk. Considered first is the problem of characterizing the amount of solar energy absorbed by the collector when it is aimed directly at the sun.

INITIAL OBJECTIVE. *To characterize the amount of solar energy absorbed by a disk which lies at the bottom of an inverted, truncated, reflecting cone aimed directly at the sun.*

When the cone is aimed directly at the sun (**Figure 1**), incoming rays hitting the cone at the same height are reflected similarly. A reflected ray remains in the same plane (the plane determined by the incoming ray and the axis of the cone) until it is either absorbed by the collecting disk or reflected outward. Rays that pierce the aperture at points equidistant from the center of the aperture are reflected in exactly the same manner. Thus, we may determine the behavior of an entire ring of incoming rays by studying the path of a single representative of that ring. Section 3 contains an analysis of the direct ray case, which leads to the accomplishment of our initial objective. Finding the solution to this first problem incorporates elements of algebra, trigonometry, and calculus.

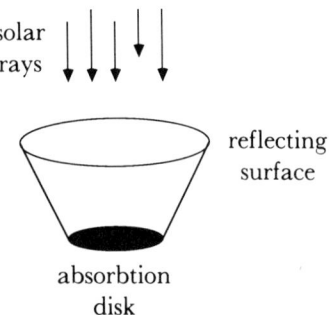

Figure 1

The initial problem leads to an investigation of the performance of the solar collector when it is not aimed directly at the sun.

ULTIMATE OBJECTIVE. *To characterize the amount of solar energy absorbed by a disk that lies at the bottom of an inverted, truncated, reflecting cone* not aimed directly at the sun.

Determining the power of the solar collector when it is not aimed directly at the sun is a more general and complex problem. Unlike the previous problem, symmetry cannot be used to reduce the problem to a two-dimensional one. Our ultimate objective requires analytic geometry and three-dimensional vector analysis. Due to the complexity of the problem, closed-form solutions will not be obtained. Numerical approximations to the solutions will be achieved, with the aid of a computational algorithm developed in section 4.

2. Fundamental Concepts

2.1 Characterization of the Reflecting Cone

An inverted, truncated cone may be specified in many ways. Consider **Figure 2** in which: R is the aperture radius of the cone; h is the height of the truncated cone; r is the base radius; H is the height of the cone; and θ is the cone angle measured from the axis of the cone.

Problem 1.
Find formulas that express
a. ψ as a function of θ;
b. H as a function of θ and R;
c. h as a function of θ, R, and r;
d. θ as a function of r, h, and R;
e. R as a function of θ, r, and h.

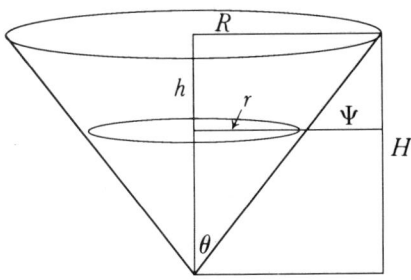

Figure 2

Note that the specification of any three of the four parameters θ, R, r, and h determines the truncated cone. Also, the value of ψ determines the value of θ, and vice versa. Which three values would you choose to determine the truncated cone?

2.2 Characterization of the Energy Carried by Solar Rays

FUNDAMENTAL ASSUMPTION 1. *Solar rays received at the earth's surface are parallel and are propagated at a constant velocity* v.

FUNDAMENTAL ASSUMPTION 2. *The amounts of solar energy contained within equal volumes in space at the earth's surface are the same.*

The first assumption is obtained by neglecting the angular size of the sun $(1/2°)$. The second follows from the fact that the sun is an *isotropic* source of light, that is, one which emits light uniformly in all directions. The two preceding assumptions lead to the following conclusion: During a given interval of time, the amounts of solar energy passing through planar areas having equal area projections on the plane normal to the direction of the rays are the same. Thus, the amount of energy passing through a given planar region during each unit of time (*power*) is proportional to the area of the region projected onto the plane normal to the direction of propagation

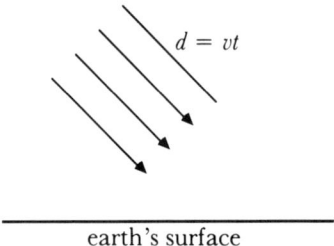

Figure 3

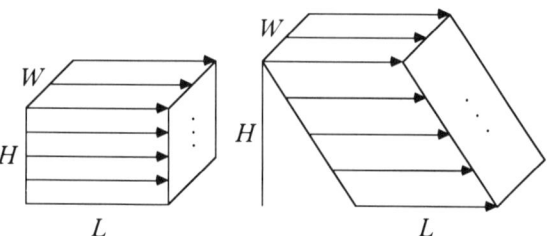

Figure 4

(*effective area*) [Tippens 1978]. The constant of proportionality is called the *intensity*, I, of the solar radiation.

Power = (Intensity)(Effective Area)
$$P = (I)(EA)$$

2.3 Power Passing through the Aperture of the Cone

The problem of computing the power absorbed by the solar collector can be solved by finding out how much of the power passing through the aperture of the cone is transmitted to the collecting disk by direct and reflected rays.

Problem 2.
What is the power of solar radiation of intensity I passing through a circular aperture of radius R which is perpendicular to the direction of propagation of the rays?

Now suppose that the direction of the solar rays makes an angle β with respect to the axis of the cone. Consider $ABCD$, a rectangular subregion of the aperture, pictured in **Figure 5**.

Problem 3.
Show that $AB'C'D$, the perpendicular projection of a rectangle $ABCD$ onto the plane perpendicular to the direction of the rays, is $lw \cos \beta$.

Problem 4.
By considering partitions of the cone's aperture composed of smaller and smaller rectangles, find the power of the radiation passing through the aperture of radius R when the direction of the solar rays makes an angle β with the axis of the cone.

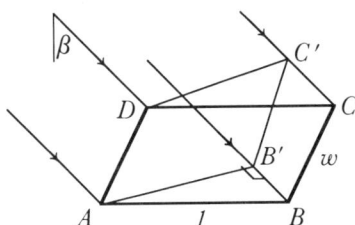

Figure 5

3. Direct Ray Analysis

Throughout this section it is assumed that the axis of the reflecting cone is parallel to the direction of the solar radiation.

3.1 Fundamental Law of Reflection

LAW OF REFLECTION. *The angle formed by an incoming ray and the normal to the reflecting surface is equal to the angle formed by the reflected ray and the normal to the reflecting surface.*

Problem 5.
Considering the law of reflection, what is the relation between β_i and β_{i+1} in **Figure 6**?

Problem 6.
If the initial incoming ray is parallel to the axis of the cone, express in terms of θ: $\beta_1, \beta_2, \ldots, \beta_i$.

Problem 7.
Use **Figure 7** to help find a criterion for the value i of the first occurrence of an outward or horizontal reflection when the initial incoming ray is parallel to the axis of the cone.

Problem 8.
a. According to the criterion found in **problem 7**, which values of θ allow only direct rays to reach the absorption disk?

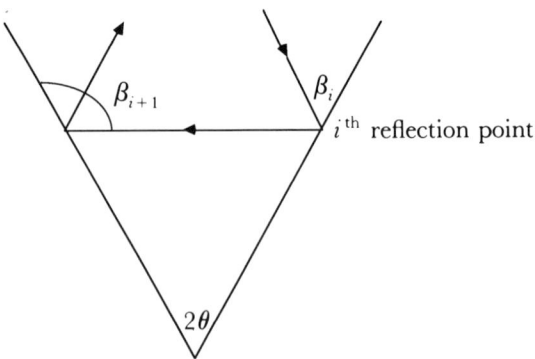

Figure 6

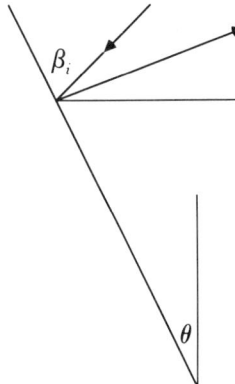

Figure 7 Criterion for outward reflection

b. Which values of θ allow only direct and first reflections to reach the disk?

c. Which values of θ allow second reflections to reach the disk?

d. What general statement can be made about the reflection of rays as the cone angle θ gets smaller?

3.2. Cone Parameters

As seen in the previous section, θ limits the maximum number of reflections a ray can undergo and still be incident on the absorption disk at the base of the reflecting cone. For a given θ, a conical reflector of any height can be built, but realistically, it should not go beyond the height at which the rays begin to be reflected out. This limiting height of the cone is also affected proportionately by the dimension of the absorption disk. We shall assume the radius of the absorbing disk, r, is 1 unit. That is, all lengths will be measured relative to the radius of the disk. The selection of the two cone parameters θ and r leaves one parameter, either R or h, to be specified. Recall that formulas relating the four quantities θ, r, R, and h were found in **Problem 1**. We shall determine R and h by fixing the number of reflections, n, that a ray can undergo and still be incident on the absorption disk. The number n must be less than or equal to n_{max}, the maximum number of reflections allowed by the selection of the cone angle θ. For example, if $\theta = 15°$, then $n \leq n_{max} = 2$. In summary, the cone will be specified by setting $r = 1$, selecting θ (which determines n_{max}), and then selecting $n \leq n_{max}$ (which determines R and h).

3.3. Radii of Reflection

For a given cone angle $\theta < 45°$, describe the ring of points of the aperture of the cone admitting solar rays that are reflected to the absorbing disk after one reflection.

Consider the incoming ray that reflects to the opposite side of the absorption disk. This will give the maximum radius, r_1, of the region under consideration.

Problem 9.
Identify the angles marked with μ in **Figure 8** and show that $r_1 = (m_1 + 1)/(m_1 - 1)$, where $m_1 = \tan(2\theta)/\tan\theta$.

Problem 10.
a. What happens for r_1 as θ decreases? **Hint:** $\tan(2\theta) = 2\tan\theta/(1 - \tan^2\theta)$.
b. What value does r_1 approach as θ approaches 0? **Hint:** Use the hint above to show that m_1 approaches 2 as θ approaches 0.
c. As r_1 approaches 1, what does θ approach? Does this make sense?

For a given cone angle $\theta < 45°$, describe the ring of points of the aperture of the cone admitting solar rays that are reflected to the absorbing disk after two reflections.

Problem 11.
a. Use **Figure 9** to show that $r_1 = [(m_1 + 1)/(m_1 - 1)]r_2$, where $m_1 = (\tan 2\theta)/\tan\theta$. Compare with **problem 9**.
b. Show that $r_2 = (m_2 + 1)/(m_2 - 1)$, where $m_2 = \tan(4\theta)/\tan\theta$.
c. As θ approaches 0, what values does r_1 approach? **Hint:** Show first that m_2 approaches 4 as θ approaches 0, by using the hint from **problem 10a**.

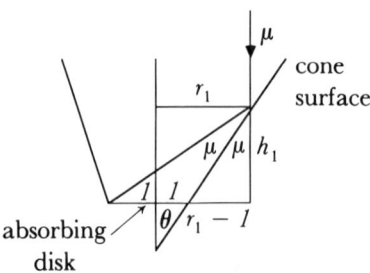

Figure 8

8

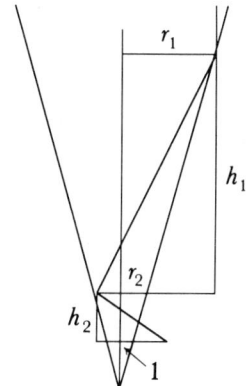

Figure 9

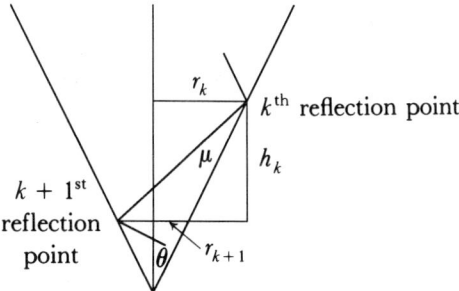

Figure 10. General relation.

Problem 12.
Consider **Figure 10**.
a. Recalling **problem 6**, determine the value of angle μ.
b. Derive the general relation

$$r_k = [(m_k + 1)/(m_k - 1)]r_{k+1}, \text{ where } m_k = \tan(2k\theta)/\tan\theta.$$

It follows from the results of **problem 12** that after n reflections

$$r_1 = [(m_1 + 1)/(m_1 - 1)][(m_2 + 1)/(m_2 - 1)] \cdots [(m_n + 1)/(m_n - 1)],$$

where

$$m_k = \tan(2k\theta)/\tan\theta, \quad k = 1, 2, \ldots, n.$$

Problem 13.
a. What does m_k approach as θ approaches 0?
b. Verify that for a fixed value k, as θ approaches $45°/k$, m_k increases without bound, but the factor $(m_k + 1)/(m_k - 1)$ approaches 1.
c. For a given n, what does r_1 approach as θ approaches 0?

Rings of Effective Area

Let $R_n = r_1$ for n reflections, $n \leq n_{\max}$. Then, by the formula preceding **problem 13**, $R_n = [(m_n + 1)/(m_n - 1)]R_{n-1}$, where $m_n = \tan(2n\theta)/\tan\theta$. The ring of contact points on the cone at which incoming rays are reflected exactly n times before being incident on the absorption disk has an effective area $EA_n = \pi(R_n^2 - R_{n-1}^2)$.

3.4 Concentrating Properties of Ideal Conical Reflectors

In an ideal solar collector, all solar radiation is reflected completely by the reflector and is absorbed completely by the collection disk. If every ray that enters the ideal collector is incident on the absorption disk, either directly or after several reflections, then the power of the collector is computed simply by multiplying the intensity of the radiation by the total effective area which is the area of the aperture of the cone. Sample calculations are given for a few special cone angles in **Table 1**.

Table 1

$\theta = 30°$.
Maximum number of reflections allowed $< 45°/30° \Rightarrow n_{\max} = 1$.

n	m_n	R_n	h_n	EA_n(times π)
0		1	0	1
1	3	2	$\sqrt{3}$	3
				Total: 4

$\theta = 22.5°$.
Maximum number of reflections allowed $< 45°/22.5° \Rightarrow n_{\max} = 1$.

n	m_n	R_n	h_n	EA_n(times π)
0		1	0	1
1	$1+\sqrt{2}$	$1+\sqrt{2}$	$2+\sqrt{2}$	$2+2\sqrt{2}$
				Total: $3+2\sqrt{2}$

10

Problem 14.
Complete a table like **Table 1** for the cone angle $\theta = 15°$.

3.5 Nonideal Reflectors

Reflectivity Factors

If the reflective properties of the cone surface are not ideal, then some of the solar radiation is absorbed rather than reflected each time a ray hits the surface. A simple model for the loss of power due to imperfect reflection is to multiply the ideal power generated by a particular region of the cone by a reflectivity constant $\Omega \leq 1$ for every time incoming rays bounce off the surface of the cone. The power collected by each ring of the cone for which the incoming solar rays are reflected n times before falling on the absorption disk is reduced by a factor of Ω^n.

Absorption Factors

Just as the reflecting surface of the cone may not be perfect, the absorption properties of the collection disk may not be ideal. A simple model of a nonideal absorbing disk analagous to the one used above for the reflecting cone is to introduce absorption factors $f_n \leq 1$. The properties of the coating used on the absorption disk may make it dependent on the angle at which reflected rays strike the collector (μ_n in **Figure 11**). For example, a simple model would be an absorption factor proportional to the sine of the entry angle, $f_n = a(\sin \mu_n)^q$, $0 < a \leq 1$, $q \geq 0$.

Problem 15.
Recalling the angle x in **Figure 11** from **problem 12a**, compute the entry angle μ_n in terms of θ.

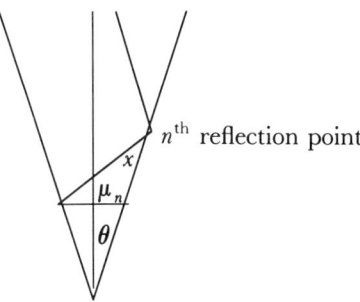

Figure 11

Table 2

$\theta = 30°$

n	EA_n (times π)	Ω^n	$\mu_n(°)$	f_n	P_n (times $I\pi$)
0	1	1.0	90	1.0	1.0
1	3	0.9	30	0.5	1.35

Power magnification (PM = Total Power/P_0): 2.35

The angle μ_n will be the same for all rays undergoing exactly n reflections before striking the collector. The power contributed by the effective area EA_n may be computed for the nonideal reflector by $P_n = I\pi EA_n \Omega^n f_n$. Table 2 was constructed with $\Omega = 0.9$ and $f_n = \sin \mu_n$.

Problem 16.
Complete tables similar to **Table 2** for $\theta = 22.5°$ and $\theta = 15°$. Use $\Omega = 0.9$ and $f_n = \sin \mu_n$.

Problem 17.
Write a computer program that computes for any arbitrarily chosen conical solar collector a table that incorporates the information found in the tables of **problems 14** and **16**. Inputs: cone angle ($0° < \theta < 45°$) and reflectivity constant ($\Omega \leq 1$). Outputs: maximum number of reflections allowed ($n_{\max}$), m_n, large radius of the cone (R_n), height of the cone (h_n), effective area (EA_n), reflectivity factor (Ω^n), entry angle (μ_n), absorption factor (f_n), power (P_n), power magnification (PM).

4. Oblique Ray Analysis

In this section it is assumed that the solar rays entering the aperture of the collector make an angle β with the axis of the reflecting cone.

4.1 Incoming Ray Description

Suppose that the solar rays entering the collector are parallel to the yz-plane and are directed in the direction of the positive y-axis.

Problem 18.
Use **Figure 12** to determine the unit vector in the direction of the solar rays.

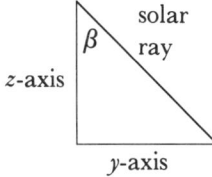

Figure 12

4.2 Partitioning of the Aperture

The solar radiation that passes through the aperture of the conical reflector either strikes the collecting disk diretly, strikes the disk after being reflected one or more times by the cone surface, or is reflected out of the cone without striking the absorption disk. Consider the region of the aperture divided into small polar rectangular subregions. For example, **Figure 13** depicts a partitioning of the aperture in which the radius of the aperture, R, is divided into five equal parts, $\Delta r = R/5$, and the 2π radians of the aperture circle are divided into eight equal angles, $\Delta \mu = 2\pi/8$. The total power of the solar collector will be computed by summing the power of the solar radiation that passes through each individual subregion of the aperture. The center point of the subregion, (r_i, μ_j), will serve as a representative of each point in the entire subregion. The path of the ray that enters the conical reflector at that point will approximate the path of any other ray that passes through a point of the subregion.

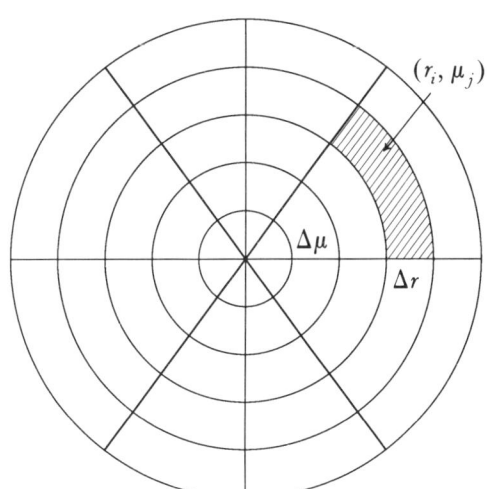

Figure 13

Problem 19.
Consider the partitioning of the aperture in **Figure 13**. Show that the shaded region with center (r_i, μ_j) has an area of $r_i \Delta\mu \Delta r$.

4.3 Computing the Power of the Collector

Recalling the results of **problems 3** and **4**, we see that the power of the solar radiation passing through a given curvilinear subregion of the aperture is given by $P = I r_i \Delta\mu \Delta r \cos \beta$. By tracing the path of a typical ray passing through the subregion, we can discover if the energy passing through the subregion is eventually absorbed by the disk; and if so, how many reflections occur before it is absorbed by the collecting disk. The power passing through the subregion may then be modified by multiplying by an appropriate factor indicating the loss of power due to imperfect reflection (Ω^n, for example), and a factor indicating the loss of power due to imperfect energy absorption of the rays by the collecting disk (for example, $f_n = \sin \mu_n$, where μ_n is the entry angle of the ray). To find the path of a given ray, an analytic description of the cone is needed.

4.4 Analytic Description of the Cone

We shall view the cone as a surface points obtained by revolving about the z-axis a line segment making an angle θ with the z-axis.

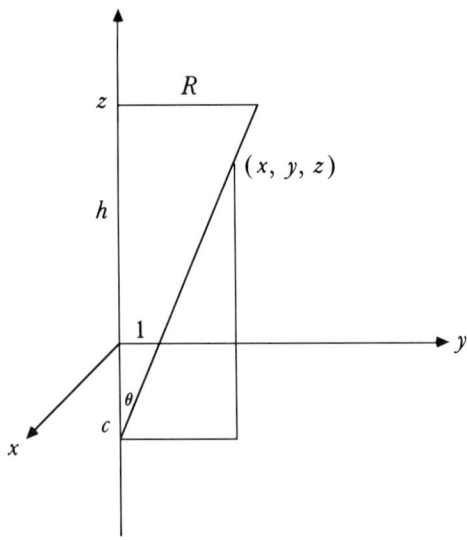

Figure 14

Problem 20.
a. Use **Figure 14** to obtain the equation for the cone $(z/c + 1)^2 - (x^2 + y^2) = 0$, where $c = \cot \theta$.
b. What additional points in space satisfy the equation given in part **a**.
c. What geometric object does the cone approximate as $c \to \infty$?

4.5 Intersection of an Incoming Ray with the Cone

Let the coordinates of a point of a ray be (x_0, y_0, z_0) and the unit vector in the direction of the ray be $\langle d_1, d_2, d_3 \rangle$. Then the coordinates of the points on the ray are given by the parametric equations $(x, y, z) = (x_0, y_0, z_0) + t(d_1, d_2, d_3)$, $t \geq 0$.

Problem 21.
Use the equation for the cone in **problem 20** to show that the point of intersection of the ray described above and the cone occurs at a value t that satisfies the quadratic equation $At^2 + 2Bt + C = 0$, where

$$A = d_3^2/c^2 - d_1^2 - d_2^2,$$

$$B = (z_0/c + 1)d_3/c - x_0 d_1 - y_0 d_2,$$

and

$$C = (z_0/c + 1)^2 - x_0^2 - y_0^2.$$

Problem 22.
Use **Figure 14** to show that $R = h/c + 1$.

Problem 23.
Let the coordinates of a point of the aperture be (x_0, y_0, h) and the components of the unit direction vector of the ray be $\langle 0, \sin \beta, -\cos \beta \rangle$. Show that for these values the constants A, B, and C of **problem 21** simplify to $A = \cos^2 \beta / c^2 - \sin^2 \beta$, $B = -R \cos \beta / c - y_0 \sin \beta$, and $C = R^2 - x_0^2 - y_0^2$.

Problem 24.
Consider A of **problem 23**. Show $A > 0$ if $0 < \beta < \theta$, $A = 0$ if $\beta = \theta$, and $A < 0$ if $\theta < \beta < \pi/2$.

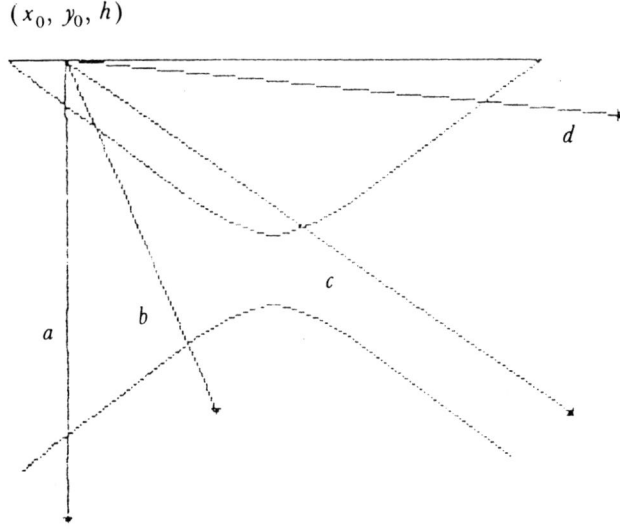

Figure 15

As a ray enters the cone in a direction parallel to the yz-plane, it is confined to the plane $x = x_0$. The intersection of this plane with the opposing cones, represented by the equation $(z/c + 1)^2 - x^2 - y^2 = 0$, is a hyperbola.

Problem 25.
Use **Figure 15** to justify the use of the formula $(-B - \sqrt{(B^2 - AC)})/A$, and $(-B + \sqrt{(B^2 - AC)})/A$, for the value of t at which the entering ray intersects the cone, whether A is positive or negative. **Hint:** Ray a represents the case $\beta = 0$, ray b the case $0 < \beta < \theta$, ray c the case $\beta = \theta$, and ray d the case $\theta < \beta < \pi/2$.

Problem 26.
When $\beta = \theta$, $A = 0$. Show that in this case the value of t at which the entering ray intersects the cone is $(R^2 - x_0^2 - y_0^2)/[2 \sin \beta (R + y_0)]$. Note that the problems with this expression arise only when $y_0 = -R$ (this is the case of the ray directed down the side of the cone).

Once the value of t is found for the intersection point, its coordinates are obtained from the equation appearing just before **problem 21**.

4.6 Intersection with the Collecting Disk

Once the intersection point with the cone is found, a check must be made to determine if the ray pierces the unit disk in the xy-plane.

Problem 27.
Using the coordinates of the intersection point with the cone, state a condition for the ray to pierce the unit disk in the xy-plane.

Problem 28.
Suppose that the ray is incident on the collecting disk. Find a formula for the entry angle μ, the angle between the xy-plane and the ray.

4.7 Inward Normal to the Cone Surface

If the ray does not strike the collecting disk, the direction of the reflected ray needs to be found. To do so, it is first necessary to find the inward normal to the cone surface.

Problem 29.
Find the unit vector $\vec{n}$ which is perpendicular to the surface of the cone and points toward the cone's interior.

4.8 Computation of the Direction of the Reflected Ray

Problem 30.
Let $\vec{d}$ be the unit vector in the direction of the incoming ray, $\vec{n}$ be the unit inward normal to the cone surface, and $\vec{b}$ be the unit vector in the direction of the reflected ray. Use **Figure 16** to show that the components of vector $\vec{b}$ may be found with the vector equation $\vec{b} = \vec{d} - 2(\vec{d} \cdot \vec{n})\vec{n}$.

Problem 31.
What is the criterion for outward reflection?

4.9 New Intersection Point

Once the direction of the reflected ray is obtained, a new point of intersection with the cone must be found. Let (x_0, y_0, z_0) now

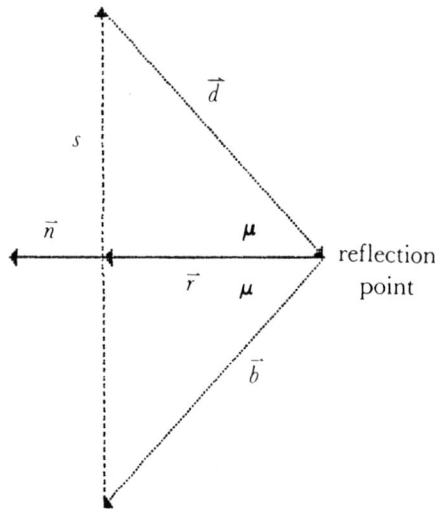

Figure 16

represent a point on the cone from which a reflected ray emanates, and let $\vec{d}$ be the unit vector in the direction of the reflected ray (the vector $\vec{b}$ computed in **problem 30**).

Problem 32.
Show that the value of the parameter t corresponding to the new point of intersection of the reflected ray with the cone is given by the equation $t = -2B/A$, where A and B are given in **problem 21**.

Once a new point of intersection with the cone has been found, its z-coordinate must be checked, as was done in section 4.6, to determine if the reflected ray is incident on the collecting disk. The algorithm outlined in Sections 4.6–4.9 is repeated, until it is discovered whether the ray is incident on the collecting disk or is reflected outwards.

4.10 Power Magnification

The total power absorbed by the collecting disk is the sum of the contributions from each of the individual polar rectangular subregions of the aperture. Each subregion contributes a power component

$$P = (\text{Solar Intensity})(\text{Effective Area})$$
$$\times (\text{Reflectivity Factor})(\text{Absorption Factor}).$$

The effective area of the polar rectangular subregion with center point (r_i, μ_j) was given in Section 4.3 by the expression $r_i \Delta r \Delta \mu \cos \beta$. As in section 3.5, we shall consider the reflectivity factor Ω^n, where Ω represents the fraction of power that is transmitted after each reflection and n is the number of reflections that the ray undergoes before hitting the collecting disk. Again, the general formula $f = a(\sin \mu)^q$, $0 < a \leq 1, q \geq 0$, will be used to model the absorption factor, where μ is the angle formed by the incident ray and the collecting disk, and the parameters a and q are selected to model the absorption characteristics of the coating on the collecting disk.

Two relative power ratios are of interest. The first ratio measures the effect that the angle of the solar rays has on the efficiency of the conical reflection. The *power efficiency at angle β* will be computed by the formula

PE_β = (power absorbed by the collector when the entering rays form an angle β with the axis of the cone)/(power absorbed by the collector when the entering rays are parallel to the axis of the cone).

The second ratio measures the effect that the reflecting cone has on the power of the collector. The *power magnification at angle β* is given by the formula

PM_β = (power absorbed by the collector when the entering rays form an angle β with the axis of the cone)/(power absorbed by the disk without the conical reflector when the entering rays form an angle β with the axis of the cone).

The denominator of this ratio is given by the expression $I\pi \cos \beta f$, where f is the absorption factor corresponding to an entry angle of $\mu = 90° - \beta$.

Table 3 summarizes the power efficiency at angle β and the power magnification at angle β for the conical reflector with the following specifications (all length measurements are given with respect to the radius of the collecting disk, which is assumed to be 1 unit): cone angle = $30°$; height of truncated cone = 1.732; radius of aperture of the cone = 2; reflectivity constant = 0.9; absorption factor = $\sin \mu$.

Table 3

$\beta(°)$	0	5	10	15	20	25	30	35	40	45	50	55	60
PE_β	1.00	0.94	0.84	0.74	0.64	0.54	0.43	0.32	0.21	0.14	0.09	0.02	0
PM_β	2.35	2.22	2.05	1.86	1.71	1.53	1.35	1.12	0.84	0.64	0.50	0.14	0

Problem 33.
Add to the computer program of **problem 17** so that it allows the radius and the height of the cone to be specified and then computes PE_β and PM_β for any angle β. Compare the performance of the 30° cone with that of a 22.5° cone of optimal height and a 15° cone of optimal height.

5. Summary

The solar concentrating properties of any conical reflector can be summarized by a table similar to **Table 3**. Yet it must be remembered that the accuracy of the numbers in the table depends on how realistically the performance of the collector is modeled. Perhaps a more complicated model of the power losses during reflection (one that uses information about the angles of reflection) should be considered. A more appropriate model of the absorption factor might also be needed. At the very least, a good model should reveal the general trends of data obtained from the testing of experimental reflectors. The importance of a study like this is that it gives the designer an idea of the relative merits of different conical reflectors without having to actually build them.

The concentrating properties of other types of solar collectors have been studied. These include: cones with cylindrical or conical targets [Cobble 1962], hemispheres with tracking absorbers [Clausing 1976], cylindrical troughs with cylindrical targets [Russell 1976], as well as truncated pyramids and hexagonal cones [Burkhardt et al. 1977], and paraboloids of revolution with cross-sectional absorption plates [Welford and Winston 1978].

6. Answers to the Problems

1. a. $\psi = 90° - \theta$.
   b. $H = R/\tan\theta$.
   c. $h = (R - r)/\tan\theta$.
   d. $\theta = \arctan[(R - r)/h]$.
   e. $R = r + h\tan\theta$.

2. $P = I\pi R^2$.

3. First show that $< BAB' = \beta$.

4. $P = I\pi R^2 \cos\beta$.

Solar Concentrating Properties of a Conical Reflector   275

5. $\beta_{i+1} = \beta_i + 2\theta$.

6. $\beta_1 = \theta, \beta_2 = 3\theta, \ldots, \beta_i = (2i-1)\theta$.

7. Ray reflects outward or horizontally after the ith reflection if $i \geq 45°/\theta$.

8. a. $\theta \geq 45°$.
   b. $22.5° \leq \theta < 45°$.
   c. $15° \leq \theta < 22.5°$.
   d. The smaller θ is, the more reflections a ray can undergo and still be absorbed.

9. $\mu = \theta$. Use $\tan 2\theta = (r_1 + 1)/h_1$ and $\tan \theta = (r_1 - 1)/h_1$.

10. a. r_1 increases as θ decreases.
    b. r_1 approaches 3 as θ approaches 0.
    c. θ approaches 45° as r_1 approaches 1. Yes, because the first reflection of a ray is almost horizontal when θ is close to 45°.

11. a. Use $\tan 2\theta = (r_1 + r_2)/h_1$ and $\tan \theta = (r_1 - r_2)/h_1$. The same multiplier was derived in **problem 9**.
    b. Use $\tan 4\theta = (r_1 + 1)h_2$ and $\tan \theta = (r_2 - 1)/h_2$.
    c. $(m_1 + 1)/(m_1 - 1)$ approaches 3 and $(m_2 + 1)/(m_2 - 1)$ approaches 5/3 as θ approaches 0. Thus, r_1 approaches 5 as θ approaches 0.

12. a. $\mu = (2k-1)\theta$.
    b. Use $\tan(2r\theta) = (r_k + r_{k+1})/h_k$ and $\tan\theta = (r_k - r_{k+1})/h_k$.

13. a. m_k approaches $2k$ as θ approaches 0.
    b. As $k\theta$ approaches 45°, $\tan(2k\theta)$ and m_k are unbounded. Thus, $(m_k + 1)/(m_k - 1)$ is approaches 1.
    c. r_1 approaches $2n + 1$ as θ approaches 0.

14. $\theta = 15°$.
    Maximum number of reflections allowed $< 45°/15° \Rightarrow n_{\max} = 2$.

n	m_n	R_n	h_n	EA_n(times π)
0		1	0	1
1	2.155	2.732	6.464	6.464
2	6.464	3.732	10.196	6.464
				Total: 12.928

15. $x = (2n-1)\theta$ implies $\mu_n = 90° - 2n\theta$.

16. $\theta = 22.5°$.

n	EA_n(times π)	Ω^n	$\mu_n(°)$	f_n	P_n(times $I\pi$)
0	1	1.0	90	1.0	1.0
1	4.828	0.9	45	0.707	3.073
				Power Magnification:	4.073

$\theta = 15°$.

n	EA_n(times π)	Ω^n	$\mu_n(°)$	f_n	P_n(times $I\pi$)
0	1	1.0	90	1.0	1.0
1	6.464	0.9	60	0.866	5.038
2	6.464	0.81	30	0.5	2.618

Power Magnification: 8.656

18. $\langle 0, \sin\beta, -\cos\beta \rangle$

19. Area $= \Delta\mu(r_i + \Delta r/2)^2/2 - \Delta\mu(r_i - \Delta r/2)^2/2$.

20. a. Note that $\cot\theta = (z + c)/\sqrt{x^2 + y^2} = c$.
    b. The points of the opposing cone which opens toward the negative z axis.
    c. A cylinder of radius 1, centered at the origin, and perpendicular to the xy plane.

21. Substitute the parametric equations for the ray into the equation for the cone obtained in **problem 20**.

22. Use $\tan\theta = 1/c = R(h + c)$.

23. Use $z_0 = h$, $d_1 = 0$, $d_2 = \sin\beta$, and $d_3 = -\cos\beta$.

24. $A = \cos^2\beta/c^2 - \sin^2\beta = \cos^2\beta(1/c^2 - \tan^2\beta) = \cos^2\beta(\tan^2\theta - \tan^2\beta)$.

25. When $\beta < \theta$ (as is the case for rays a and b), **Figure 15** shows that the quadratic equation has two positive roots, the smaller of which we desire. Since $A > 0$ (by **problem 24**), $-B/A - \sqrt{B^2 - AC}/A < -B/A + \sqrt{B^2 - AC}/A$. When $\beta > \theta$ (as is the case of ray d), **Figure 15** shows that the quadratic equation has one positive and one negative root. The fact that A is now negative reverses the above inequality, showing that $-B/A - \sqrt{B^2 - AC}/A$ is the positive root that we desire.

26. $t = -C/(2B) = (R^2 - x_0^2 - y_0^2)/\{2[R(\cos\beta)/c + y_0\sin\beta]\} = (R^2 - x_0^2 - y_0^2)/[(R + y_0)\sin\beta]$, because $c = \cot\beta$.

27. $z \leq 0$ or $x^2 + y^2 \leq 1$.

28. $\mu = \sin^{-1}(-d_3)$.

29. The gradient of the left hand side of the cone equation is normal to the cone surface. Since $\vec{g} = \langle -2x, -2y, 2(z/c + 1)/c \rangle$ points upward, we know that this is an inward normal. Since $z/c + 1 = \sqrt{x^2 + y^2}$, the direction of $\vec{g}$ simplifies to $\langle -x, -y, \sqrt{x^2 + y^2}/c \rangle$. Normalizing yields $\vec{n} = \cos\theta \langle -x/\sqrt{x^2 + y^2}, -y/\sqrt{x^2 + y^2}, 1/c \rangle$.

30. Use $\vec{b} = \vec{r} - \vec{s}$ and $-\vec{d} = \vec{r} + \vec{s}$ to show $\vec{b} = \vec{d} + 2\vec{r}$. But, $\vec{r} = -(\vec{d}\cdot\vec{n})\vec{n}$, yielding $\vec{b} = \vec{d} - 2(\vec{d}\cdot\vec{n})\vec{n}$.

31. $b_3 \geq 0$.

32. Since the point of origin of the ray $\langle x_0, y_0, z_0 \rangle$ satisfies the equation of the cone, $C = 0$.

References

Burkhard, Donald G., David L. Shealy, and George Strobel. 1977. Comparison of the solar concentrating properties of truncated hexagonal, pyramidal and circular cones. *SPIE Optics Applied to Solar Energy Conversion* 114:67–94.

Clausing, A.M. 1976. Optical and thermal characteristics of a solar collector with a stationary sperical reflector and a tracking absorber. *SPIE Optics in Solar Energy Utilization II* 85:128–138.

Cobble, M.H. 1963. Analysis of a conical solar concentrator. *Solar Energy* 7(2):75–78.

Lee, John Francis, and Francis Weston Sears. 1963. *Thermodynamics* Reading, MA: Addison-Wesley.

Meinel, Aden B., and Marjorie P. Meinel. 1976. *Applied Solar Energy, An Introduction* Reading, MA: Addison-Wesley.

Russell, John L., Jr. 1976. Principles of the fixed mirror solar concentrator. *SPIE Optics in Solar Energy Utilization II* 85:139–145.

Senft, J.R. 1986. A solar Ringbom Stirling engine. *Proceedings of the 21st Intersociety Energy Conservation Engineering Conference*. San Diego: American Chemical Society.

Tippens, Paul E. 1978. *Applied Physics*. New York: McGraw-Hill.

Welford, W.T., and R. Winston. 1978. *The Optics of Nonimaging Concentrators: Light and Solar Energy*. New York: Academic Press.

UMAP
Module 681

Modules in Undergraduate Mathematics and its Applications

Simple Mortality Functions

Ho Kuen Ng

Published in cooperation with the Society for Industrial and Applied Mathematics, the Mathematical Association of America, the National Council of Teachers of Mathematics, the American Mathematical Association of Two-Year Colleges, The Institute of Management Sciences, and the American Statistical Association.

INTERMODULAR DESCRIPTION SHEET:	UMAP Unit 681
TITLE:	SIMPLE MORTALITY FUNCTIONS
AUTHOR:	Ho Kuen Ng Department of Mathematics and Computer Science San Jose State University San Jose, CA 95192
MATHEMATICAL FIELD:	Applied mathematics
APPLICATION FIELD:	Mortality studies, actuarial science
TARGET AUDIENCE:	Students in courses on differential equations or probability
ABSTRACT:	This unit introduces some simple functions which describe human mortality. The derivations and limitations are discussed.
PREREQUISITES:	Acquaintance with differential equations and probability

© Copyright 1987 by COMAP, Inc. All rights reserved.

COMAP, 60 Lowell Street, Arlington, MA 02174 (617) 641-2600

Simple Mortality Functions

Ho Kuen Ng
Department of Mathematics and
Computer Science
San Jose State University
San Jose, CA 95192

Table of Contents

1. INTRODUCTION 1
2. THE SIMPLEST MODEL 1
3. FORCE OF MORTALITY 2
4. GOMPERTZ'S LAW 4
5. MAKEHAM'S LAW 5
6. CONCLUSION 7
7. ANSWERS TO EXERCISES 8
8. REFERENCES 8
9. APPENDIX 9

Modules and Monographs in Undergraduate Mathematics and Its Applications (UMAP) Project

The goal of UMAP was to develop, through a community of users and developers, a system of instructional modules in undergraduate mathematics and its applications to be used to supplement existing courses and from which complete courses may eventually be built.

The Project was guided by a National Advisory Board of mathematicians, scientists, and educators. UMAP was funded by a grant from the National Science Foundation and is now supported by the Consortium for Mathematics and Its Applications (COMAP), Inc., a nonprofit corporation engaged in research and development in mathematics education.

COMAP Staff

Paul J. Campbell	Editor
Solomon A. Garfunkel	Executive Director, COMAP
Laurie W. Aragon	Business Development Manager
Philip A. McGaw	Production Manager
Nancy Hawley	Copy Editor
Annemarie S. Morgan	Administrative Assistant
John Gately	Distribution

1. Introduction

Throughout history many people have proposed models for the mortality rate of humans. Besides just pure curiosity and seeking to understand more about life, there are practical applications for studying mortality models. Population projection, social security program projection, and annuity and insurance premiums are just a few examples.

Simple and well-behaved functions were especially useful in the days when computers were not as popular. Even now, it is much easier to study a simple function that approximates well, rather than an entire table with hundreds of numbers. This is especially true when we need to study functions depending on the mortality of more than one life. This article describes some of the simple functions that have proven useful in many situations.

"... simple models of behavior are even more important now than they were before computers."

In fact, simple models of behavior are even more important now than they were before computers. They serve as guides to thinking, both in deciding what to compute, and afterward in interpreting the computed results. What can be accounted for by a simple model is explained in vain by one more complex.

2. The Simplest Model

We imagine we have a large population of l_0 newborn babies. Letting l_x represent the number surviving to at least age x, and X be the random variable representing the lifetime of an individual, we see at once that $\Pr(X \geq x) = l_x/l_0$. (We assume that our population is large enough that the preceding result is true for practical purposes.) It is clear that l_x must be a decreasing function of x.

In 1725 Abraham de Moivre introduced the first function to express mortality. His formula was

$$l_x = \begin{cases} l_0 \cdot \dfrac{(86-x)}{86} & \text{for } 0 \leq x \leq 86 \\ 0 & \text{for } x > 86 \end{cases}.$$

In other words, he assumed that 86 was the highest attainable age, and that l_x decreased linearly in x. The formula was later generalized by others to

$$l_x = \begin{cases} l_0 \cdot \dfrac{(\omega - x)}{\omega} & \text{for } 0 \leq x \leq \omega \\ 0 & \text{for } x > \omega \end{cases},$$

where ω is the highest attainable age, usually taken to be approximately 100. This ω is called the *limiting age*.

Of course, we do not have an absolute age by which all people must die. We must remember that we are only interested in simple functions that approximate human mortality. There is no way that we can know the "true" law of mortality, if there is such a law at all.

Exercises

1. Let $d_x = l_x - l_{x+1}$, the number of people who survive to at least age x but die before age $x + 1$. Show that under de Moivre's model d_x is constant.

2. Show that if l_x is quadratic in x, then d_x forms an arithmetic progression.

3. Let $l_x = a + bc^x$, where a, b, and c are constants. Show that d_x forms a geometric progression.

3. Force of Mortality

Let X be a continuous random variable, with density function f and distribution function F, which takes only positive values. (Recall that $F' = f$.) We interpret X as the lifetime of a device. The hazard rate μ of the device is defined by $\mu(x) = f(x)/(1 - F(x))$. It is the instantaneous rate of failure at time x, given that the device is still working then. By solving the easy differential equation, $\mu = F'/(1 - F)$, i.e., $F' + \mu F = \mu$, we have

$$F(x) = 1 - e^{-\int_0^x \mu(y)\, dy}.$$

Let us now interpret X as the lifetime of a human being. Thus $\mu(x)$ is the instantaneous rate of dying, given that a person is still living at age x. In other words, if Δx is a small time interval, then $\mu(x)\Delta x$ is the probability that an x-year old will die before reaching age $x + \Delta x$. Under such an interpretation, $\mu(x)$ is called in the literature the *force of mortality* at age x. The quantity x is usually written as a subscript, so that the force of mortality at age x is denoted by μ_x.

Let us now calculate formulas for F and f by applying the generalized version of the de Moivre formula. We get

$$F(x) = P(X \le x) = 1 - P(X \ge x)$$
$$= 1 - l_x/l_0 = 1 - \frac{\omega - x}{\omega} = \frac{x}{\omega},$$

so that

$$f(x) = F'(x) = \frac{1}{\omega}.$$

Thus $\mu_x = 1/(\omega - x)$. We show the graph of this function in **Figure 1**.

Experience, however, tells us that the curve of the force of human mortality resembles **Figure 2**. (See Appendix.)

Newborn babies are weak and susceptible to many different kinds of diseases; they grow stronger as they grow older. The curve is almost flat between ages 10 and 40. Then old age comes along and the curve begins increasing, more and more rapidly as the end of the life span is approached. Such a curve is generally called a *bathtub curve*, apparently from its shape.

Usually the parts of childhood and very advanced ages are the most difficult to estimate and model. In fact, these parts change rather rapidly as medical and other branches of science continue

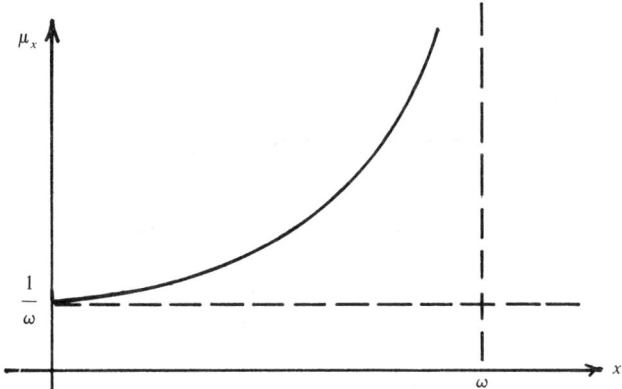

Figure 1. De Moivre's force of mortality curve.

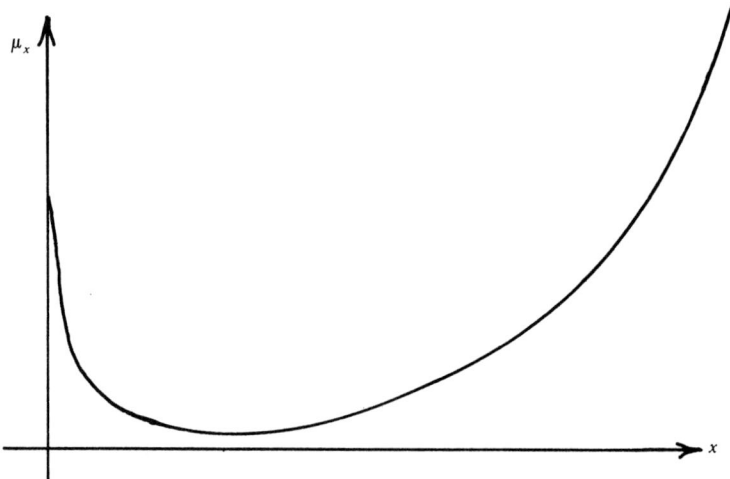

Figure 2. Empirical force of mortality curve.

their advances. As a practical matter, most studies concentrate on the part of the curve after the childhood ages, and up to ages about 70. The other parts of the mortality curve are usually determined by interpolating empirical data instead of using a specific model.

Exercises

4. Show that the hazard rate is the instantaneous rate of failure, as explained in this section.

5. Prove that

$$F(x) = 1 - e^{-\int_0^x \mu(y)\,dy}.$$

4. Gompertz's Law

In 1825 Benjamin Gompertz introduced a more sophisticated model [1825]. He reasoned as follows: Each person has a natural resistance to death, and such resistance decreases at a rate proportional to itself as the person ages. This is because in small intervals of equal lengths, equal portions of this resistance are lost. As μ_x is the instantaneous rate of death, Gompertz used $1/\mu_x$ to represent the resistance power to death in his model. Thus we obtain the differential equation

$$\frac{d}{dx}\left(\frac{1}{\mu_x}\right) = -\alpha \frac{1}{\mu_x},$$

where α is a non-negative constant. The equation can be solved as follows.

$$\frac{\dfrac{d}{dx}\left(\dfrac{1}{\mu_x}\right)}{\left(\dfrac{1}{\mu_x}\right)} = -\alpha$$

$$\log\left(\frac{1}{\mu_x}\right) = -\alpha x + \beta,$$

where β is a constant,

$$\frac{1}{\mu_x} = e^{-\alpha x} e^{\beta}$$

$$\mu_x = \gamma e^{\alpha x}$$
$$= \gamma c^x,$$

where $\gamma = e^{-\beta}$, and $c = e^{\alpha}$ are constants. To obtain l_x, observe that

$$\int_0^x \mu_y \, dy = \int_0^x \gamma c^y \, dy$$

$$= \frac{\gamma c^y}{\log c}\bigg|_0^x \quad \text{in the case when } c \neq 1,$$

$$= \frac{\gamma c^x}{\log c} - \frac{\gamma}{\log c}$$

$$= -(c^x - 1) \log g, \text{ where } \log g = -\frac{\gamma}{\log c},$$

$$= -\log g^{c^x - 1}.$$

Thus $l_x = l_0 e^{-\int_0^x \mu_y \, dy} = l_0 g^{c^x - 1}$.

The preceding formula is usually written as $l_x = k g^{c^x}$, where $k = l_0/g$.

Exercises

6. Show that under Gompertz's law, μ_x forms a geometric progression.

7. Prove that if $c = 1$ in Gompertz's law, the lifetime of an individual is an exponential random variable.

8. Show that the constants in Gompertz's law satisfy $\gamma > 0$, $c \geq 1$, and $g < 1$.

5. Makeham's Law

In 1860 Makeham carried the idea of Gompertz one step further [1860]. Gompertz, besides deriving the formula in the previous section, suggested that death might be caused by two factors, one being a constant force independent of age, and the other factor being opposed by a resistance which decreases at a rate proportional to itself. Makeham made use of this idea and derived a formula that generalizes Gompertz's law. The first factor in Gompertz's idea is called the chance or accident component, to which all ages are equally susceptible. The second factor is called the age component. Using the arguments in the previous section, we conclude that $\mu_x = \delta + \gamma c^x$, where δ, γ and c are constants.

$$\int_0^x \mu_y \, dy = \int_0^x (\delta + \gamma c^y) \, dy$$

$$= \delta x - \log g^{c^x - 1}$$

$$= -\log s^x - \log g^{c^x - 1},$$

where $s = e^{-\delta}$ is a constant,

$$= -\log s^x g^{c^x - 1}.$$

Thus $l_x = l_0 e^{-\int_0^x \mu_y \, dy} = l_0 s^x g^{c^x - 1}$.

In the literature, Makeham's law is usually written as $l_x = k s^x g^{c^x}$, where $k = l_0/g$.

Observe that with $\delta = 0$ (or $s = 1$), Makeham's law reduces to Gompertz's law.

Studies have shown that, with appropriate choices of the constants, the above equation does model empirical experience fairly well in the middle part of the mortality curve, from ages of about 10 to 60. A more accurate representation is to divide the entire age axis into shorter age intervals and apply Makeham's law to obtain different sets of constants for different intervals. This way, we can use Makeham's law to model a larger part of the life span.

Exercises

9. Prove that under Makeham's law, the first difference of μ_x forms a geometric progression.

10. Prove that the constants in Makeham's law satisfy $\gamma > 0$, $c \geq 1$, $g < 1$, $\delta \geq -\gamma$, and $s > 0$.

11. Prove that if $c = 1$ in Makeham's law, the lifetime of an individual is an exponential random variable.

6. Conclusion

Despite the fact that computing equipment has become more abundant, simple functions to describe the mortality of human beings are still an important topic. They still serve as guides in determining what to expect and in interpreting actual results. Nevertheless, we should be constantly questioning the validity of our models as new data and trends develop over time. Also, by looking at what people did in the last century, we can appreciate the insights and ingenuity that they had in simplifying a complicated mathematical problem to something more tractable.

Mathematicians who study human mortality are not fortune-tellers. They are not trying to predict how long a person will live. They are only supplying models that approximate the rate of deaths in a population. Furthermore, the mortality functions discussed in this module are only simple functions, based on plausible models, which approximate what we have been observing. They are not laws of nature that mortality must follow. It is this imprecision that makes such studies a mixture of art and science. It is also this lack of exact understanding that provides the driving force behind our quest for more knowledge.

The models we have seen are still used in models for problems of population dynamics (not just of humans!) and growth and decay. In particular, the reader can find uses of Gompertz's model in [Braun 1983, 50; Zill 1986, 111].

7. Answers to Exercises

1. $d_x = l_0/\omega$.

2. Let $l_x = ax^2 + bx + c$.
   Then $d_x = -2ax - a - b$.

3. $d_x = b(1-c)c^x$.

4. P (failure before $x + \Delta x$ | working at x)
   $= P$ (failure between x and $x + \Delta x$)/P (working at x)
   $= f(x)\Delta x/(1 - F(x))$.

5. Solve the differential equation $F' + \mu F = \mu$, with the boundary condition $F(0) = 0$.

6. An easy verification.

7. $\mu_x = $ constant. Use exercise 5.

8. As $\alpha \geq 0$, $c = e^\alpha \geq 1$.
   $\gamma = e^{-\beta} > 0$.
   As $\gamma > 0$ and $\log c > 0$, $\log g < 0$.

9. $\Delta \mu_x = \mu_{x+1} - \mu_x = \gamma(c-1)c^x$.

10. $\mu_0 = \delta + \gamma \geq 0$, $s = e^{-\delta} > 0$.

11. $\mu_x = $ constant. Use exercise 5.

8. References

Braun, M. 1983. *Differential Equations and Their Applications: An Introduction to Applied Mathematics*. 3rd ed. New York: Springer-Verlag.

Elston, J. S. 1923. Survey of mathematical formulas that have been used to express a law of mortality. *Record of the American Institute of Actuaries* 12:66–95.

Gompertz, B. 1825. On the nature of the function expressive of the law of human mortality. *Philosophical Transactions of the Royal Society of London*.

Jordan, C. W. 1982. *Life Contingencies*. Chicago: Society of Actuaries.

Makeham, W. M. 1860. On the law of mortality, and the construction of annuity tables. *Journal of the Institute of Actuaries* 8.

Myers, R. J., and F. R. Bayo, 1985. United States life tables for 1979–81. *Transactions of the Society of Actuaries* 37:303–344.

Zill, Dennis G. 1986. *Differential Equations with Boundary Value Problems.* Boston: Prindle, Weber & Schmidt.

9. Appendix

In this appendix we exhibit a real mortality table, one of the U.S. Life Tables for 1979–81. Mortality tables are prepared every 10 years in the United States. We will only look at one of the tables here, the table for the mortality rates of the entire U.S. population [Myers and Bayo 1985].

In this module, the function we are mainly interested in is the force of mortality, which is the instantaneous rate of dying. We study

Table 1.

Mortality Rates of the United States, 1979–81

Age	Mortality rate per 100,000
0	1,260
1	93
5	37
10	20
15	69
20	120
25	132
30	133
35	159
40	232
45	366
50	589
55	902
60	1,368
65	2,059
70	3,052
75	4,507
80	6,882
85	10,725
90	15,868
95	22,976
100	29,120
105	33,539

this function because it is the easiest one to work with mathematically. However, in mortality tables usually the mortality rates are listed, instead of the force of mortality. The mortality rate at age x is defined as the probability that an x-year old will die before attaining age $x + 1$. In the U.S. Tables the mortality rates are represented as number of deaths per 100,000. Tables are constructed using these rates because they are the easiest to measure in a census. Although the force of mortality and the mortality rates are different functions, their shapes are generally very similar. We exhibit the data in **Table 1**. Interested readers can sketch the corresponding graph and see that its shape is similar to **Figure 2**.

About the Author

Ho Kuen Ng received his B.S. from the University of Hong Kong and his Ph.D. degree from the University of California, Berkeley. In addition to teaching at San Jose State University, he spends some of his time doing actuarial work. His major interests are algebra, operations research, and actuarial science.

UMAP Module 684

Modules in Undergraduate Mathematics and its Applications

Linear Programming Via Elementary Matrices

Helen Wang

Published in cooperation with the Society for Industrial and Applied Mathematics, the Mathematical Association of America, the National Council of Teachers of Mathematics, the American Mathematical Association of Two-Year Colleges, The Institute of Management Sciences, and the American Statistical Association.

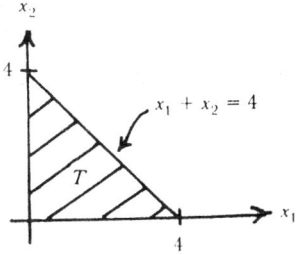

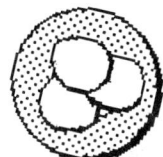

INTERMODULAR DESCRIPTION SHEET:	UMAP Unit 684
TITLE:	Linear Programming Via Elementary Matrices
AUTHOR:	Helen Wang Department of Mathematics Wellesley College Wellesley, MA 02181
MATHEMATICAL FIELD:	Linear algebra
TARGET AUDIENCE:	Students studying matrix algebra
PREREQUISITE SKILLS:	Familiarity with matrices and matrix operations; knowledge of elementary row operations is helpful but not required.
OUTPUT SKILLS:	1) Basic knowledge of linear programming. 2) Understanding of and ability to use, from a matrix point of view, the simplex method for solving linear programming problems. 3) Understanding of elementary row operations as multiplication by nonsingular elementary matrices.
ABSTRACT:	This module provides a short matrix-based introduction to the simplex method, enabling a student familiar with matrix algebra (but not necessarily with row reduction techniques) to learn how to use the simplex method to solve a linear programming problem. Several basic references are supplied for the reader interested in exploring the many applications of linear programming as well as other approaches to the simplex method.

© Copyright 1988 by COMAP, Inc. All rights reserved.

COMAP, 60 Lowell Street, Arlington, MA 02174 (617) 641-2600

Linear Programming Via Elementary Matrices

Helen Wang
Department of Mathematics
Wellesley College
Wellesley, MA 02181

Table of Contents

1. LINEAR PROGRAMMING PROBLEMS 1
   1.1. What They Are 1
   1.2. Uses 3
   1.3. A Nutrition Problem 3
2. PREPARING TO SOLVE THE NUTRITION PROBLEM 5
   2.1. Slack Variables 6
   2.2. Elementary Matrices and Elementary Row Operations . . 8
3. THE SIMPLEX METHOD 8
   3.1. Motivation and Discussion 8
   3.2. Summary 10
4. FURTHER DISCUSSION 11
5. REFERENCES 12
6. ANSWERS TO EXERCISES 13

Modules and Monographs in Undergraduate Mathematics and Its Applications (UMAP) Project

The goal of UMAP was to develop, through a community of users and developers, a system of instructional modules in undergraduate mathematics and its applications to be used to supplement existing courses and from which complete courses may eventually be built.

The Project was guided by a National Advisory Board of mathematicians, scientists, and educators. UMAP was funded by a grant from the National Science Foundation and is now supported by the Consortium for Mathematics and Its Applications (COMAP), Inc., a non-profit corporation engaged in research and development in mathematics education.

COMAP Staff

Paul J. Campbell	Editor
Solomon A. Garfunkel	Executive Director, COMAP
Laurie W. Aragon	Business Development Manager
Philip A. McGaw	Production Manager
Theresa Cronin	Copy Editor
Annemarie S. Morgan	Administrative Assistant
John Gately	Distribution

1. Linear Programming Problems

1.1. What They Are

A linear programming problem consists of optimizing (maximizing or minimizing) a linear function of one or more variables, subject to one or more linear constraints on the variables. For example,

Find the minimum value of
$$c = 15 + x_1 - 3x_2$$
subject to
$$x_1 + x_2 \leq 4$$
$$x_1 \geq 0$$
$$x_2 \geq 0 \tag{1}$$

is a linear programming problem. We may cast the problem in geometric terms: We wish to find the minimum height of the surface $f(x_1, x_2) = 15 + x_1 - 3x_2$ above the triangle T bounded by $x_1 + x_2 = 4$, $x_1 = 0$, and $x_2 = 0$ (**Figure 1**). Hence, if we envision placing a ball on that portion of the plane $z = 15 + x_1 - 3x_2$ lying above the triangle T, then problem (1) is to determine the lowest point where the ball will settle under gravity in the negative z direction (**Figure 2**). We observe that the minimum value of c is 3, which occurs at $x_1 = 0$, $x_2 = 4$. This solution also makes sense algebraically, since to minimize $c = 15 + x_1 - 3x_2$ we need to make x_1 as small as possible and x_2 as large as possible.

Exercises
Solve the following linear programming problems by geometric and/or algebraic inspection, if possible.
1. Find the maximum value of
$$c = 15 + x_1 - 3x_2$$
subject to $x_1 + x_2 \leq 4$
$$x_1 \geq 0$$
$$x_2 \geq 0.$$

2. Find the minimum value of
$$c = 21 - 2x_1 + 5x_2$$

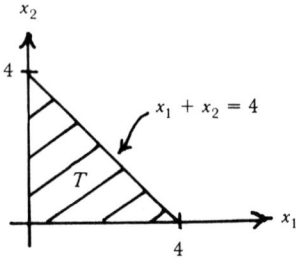

Figure 1. Constraint set for problem (1).

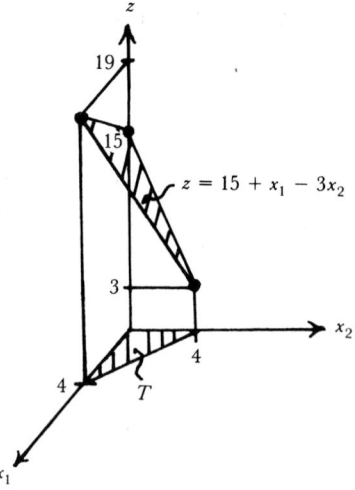

Figure 2. The triangular portion of the plane $z = 15 + x_1 - 3x_2$ lying above the constraint triangle T.

subject to $x_1 + 3x_2 \leq 9$

$x_1 \geq 0$

$x_2 \geq 0$.

3. Find the maximum value of

$$c = 21 - 2x_1 + 5x_2$$

subject to $x_1 + 3x_2 \leq 9$

$x_1 \geq 0$

$x_2 \geq 0$.

4. Find the minimum value of
$$c = 17 + 3x_1 + 5x_2$$
subject to $x_1 \geq 0$, $x_2 \geq 0$.

5. Find the maximum value of
$$c = 17 + 3x_1 + 5x_2$$
subject to $x_1 \geq 0$, $x_2 \geq 0$.

6. Find the minimum value of
$$c = 17 + 3x_1 + 5x_2$$
subject to $x_1 + x_2 \leq -1$
$$x_1 \geq 0$$
$$x_2 \geq 0.$$

1.2. Uses

"Linear programming problems have a wide variety of applications...."

Linear programming problems have a wide variety of applications, particularly in industry, including minimizing a (linear) cost function or maximizing a (linear) profit function, subject to (linear) constraints, such as warehouse capacities, budget constraints, or work force schedules. For an excellent historical introduction including applications, see [Dantzig 1963, Chapters 1, 2, 3, 15 and 17].

In the next sections we will examine carefully one specific application of a linear programming problem. We will solve a linear programming maximization problem, concentrating on matrix algebra techniques rather than on geometric methods.

1.3. A Nutrition Problem

This problem consists of determining the amounts of various foods to eat to maximize the total nutrient (e.g., protein) content in one's diet, subject to budget and calorie constraints. Let's assume there are three foods in one's daily diet, and let x_1, x_2, and x_3 denote the number of ounces of each. Suppose there are 10 units of protein per ounce of food 1, 7 units of protein per ounce of food 2, and 8 units of protein per ounce of food 3. Then the total daily amount of protein from these foods is

$$c = 10x_1 + 7x_2 + 8x_3.$$

3

We wish to maximize c, subject to budget and calorie constraints. If our daily budget is $1.20, and food 1 costs $0.20 per ounce, food 2 costs $0.10 per ounce, and food 3 costs $0.30 per ounce, then the daily budget constraint is

$$(.2)x_1 + (.1)x_2 + (.3)x_3 \leq 1.2,$$

or

$$2x_1 + x_2 + 3x_3 \leq 12.$$

Also, suppose we wish to restrict the daily calorie intake in the diet to no more than 1500 calories. If food 1 contains 200 calories per ounce, food 2 contains 200 calories per ounce and food 3 contains 100 calories per ounce, then this constraint is

$$200x_1 + 200x_2 + 100x_3 \leq 1500,$$

or

$$2x_1 + 2x_2 + x_3 \leq 15.$$

Of course, the number of units of each food must be nonnegative, so our problem is

Find the maximum value of
$$c = 10x_1 + 7x_2 + 8x_3$$
subject to
$$2x_1 + x_2 + 3x_3 \leq 12$$
$$2x_1 + 2x_2 + x_3 \leq 15$$
$$x_1 \geq 0, \quad x_2 \geq 0, \quad x_3 \geq 0. \tag{2}$$

Here geometric and/or algebraic inspection does not readily yield an answer; we will learn a technique called the *simplex method* for finding a solution.

Exercise
7. Write the following problem as a linear programming problem:
   A bakery sells two kinds of cookies, sugar cookies and gingerbread cookies. It takes 1/2 hour to make a pound of sugar cookies, 1 hour to make a pound of gingerbread cookies, and the bakery operates 12 hours per day. Only one type of cookie can be made at a time. Also, the bakery is able to sell no more than 20 pounds of cookies per day. If the bakery makes a profit of 30¢ per pound on the sugar cookies and 50¢ per pound on the gingerbread

cookies, how many pounds of each should the bakery make each day in order to maximize profits? (Do not solve yet. See **Exercise 19**.)

2. Preparing to Solve the Nutrition Problem

2.1. Slack Variables

Before we introduce the simplex method for solving linear programming problems, it will be helpful to be able to convert all constraints to either equality constraints or to nonnegativity constraints. This can be done by introducing new variables to take up the "slack" in each inequality constraint; hence the name *slack variable*. For example, consider the inequality constraint

$$2x_1 + x_2 + 3x_3 \leq 12. \tag{3}$$

We introduce a new nonnegative variable x_4 and consider the equality constraint

$$2x_1 + x_2 + 3x_3 + x_4 = 12. \tag{4}$$

If x_1, x_2, and x_3 satisfy (3), then (4) is satisfied with

$$x_4 = 12 - (2x_1 + x_2 + 3x_3) \geq 0.$$

Conversely, if x_1, x_2, x_3 and x_4 satisfy (4) with $x_4 \geq 0$, then x_1, x_2, and x_3 satisfy (3). Similarly, we can replace the inequality constraint $2x_1 + 2x_2 + x_3 \leq 15$ with the equality constraint $2x_1 + 2x_2 + x_3 + x_5 = 15$, where the new slack variable x_5 must be nonnegative. Consequently, by using two nonnegative slack variables x_4 and x_5, the original nutrition problem (2), which had two inequality constraints and three nonnegative variables, becomes

Find the maximum value of
$$c = 10x_1 + 7x_2 + 8x_3$$
subject to
$$2x_1 + x_2 + 3x_3 + x_4 \qquad = 12$$
$$2x_1 + 2x_2 + x_3 \qquad + x_5 = 15$$
$$x_1, x_2, x_3, x_4, x_5 \geq 0, \tag{5}$$

with two equality constraints and five nonnegative variables.

We are almost ready to apply the simplex method of solution to form (5) of the nutrition problem; we will complete our preparation with the introduction in the next section of two special square matrices called *elementary matrices*.

Exercises
Introduce appropriate slack variables to change the following inequality constraints to equality constraints with nonnegative constraints.

8. $3x_1 + x_2 \leq 7$.

9. $x_1 + 7x_2 \geq 5$.

2.2. Elementary Matrices and Elementary Row Operations

We begin by writing problem (5) in matrix notation.

Maximize c subject to

$$\begin{bmatrix} 2 & 1 & 3 & 1 & 0 \\ 2 & 2 & 1 & 0 & 1 \\ 10 & 7 & 8 & 0 & 0 \end{bmatrix} \begin{bmatrix} x_1 \\ x_2 \\ x_3 \\ x_4 \\ x_5 \end{bmatrix} = \begin{bmatrix} 12 \\ 15 \\ c \end{bmatrix} \quad (6)$$

and $x_1, x_2, x_3, x_4, x_5 \geq 0$.

Our goal is to rewrite (6) in an equivalent form that is easier to solve. We will use a matrix technique. (If you have learned to solve systems of linear equations by row reduction, the techniques which follow may remind you of elementary row operations and the Gauss-Jordan procedure. However, such knowledge is not necessary here.)

First, if we left-multiply both sides of the matrix equation in (6) by the same nonsingular 3×3 matrix, then the resulting matrix equation will be equivalent to the original one. (Recall that a square matrix is *nonsingular*, or *invertible*, if it has an inverse under matrix multiplication.) Consider the effect of multiplying both sides of the matrix equation in (6) on the left by the matrix

$$\begin{bmatrix} 1/2 & 0 & 0 \\ 0 & 1 & 0 \\ 0 & 0 & 1 \end{bmatrix}.$$

This operation replaces row 1 of the matrix on each side of (6) with

1/2 of row 1 (see Exercise 10):

$$\begin{bmatrix} 1 & 1/2 & 3/2 & 1/2 & 0 \\ 2 & 2 & 1 & 0 & 1 \\ 10 & 7 & 8 & 0 & 0 \end{bmatrix} \begin{bmatrix} x_1 \\ x_2 \\ x_3 \\ x_4 \\ x_5 \end{bmatrix} = \begin{bmatrix} 6 \\ 15 \\ c \end{bmatrix}. \tag{7}$$

In general, if $A(i, r)$ is the 3×3 nonsingular matrix that is the 3×3 identity matrix with the ith diagonal element replaced by the nonzero constant r, then left-multiplication by $A(i, r)$ of a $3 \times n$ matrix (n can be any positive integer) replaces row i of that matrix with r times row i (see **Exercises 11–13**).

Now we will determine the effect of left-multiplying both sides of the matrix equation in (7) by

$$\begin{bmatrix} 1 & 0 & 0 \\ -2 & 1 & 0 \\ 0 & 0 & 1 \end{bmatrix}.$$

This replaces row 2 of the matrix on each side of (7) with row 2 plus -2 times row 1 (see Exercise 14):

$$\begin{bmatrix} 1 & 1/2 & 3/2 & 1/2 & 0 \\ 0 & 1 & -2 & -1 & 1 \\ 10 & 7 & 8 & 0 & 0 \end{bmatrix} \begin{bmatrix} x_1 \\ x_2 \\ x_3 \\ x_4 \\ x_5 \end{bmatrix} = \begin{bmatrix} 6 \\ 3 \\ c \end{bmatrix}. \tag{8}$$

In general, if $B(i, j, r)$ is the 3×3 nonsingular matrix which is the 3×3 identity matrix with the element in row i and column j ($i \neq j$) replaced by r, then left-multiplication by $B(i, j, r)$ of a $3 \times n$ matrix (n can be any positive integer) replaces row i of that matrix with row i plus r times row j (see **Exercises 15–17**).

Matrices of the form $A(i, r)$ and $B(i, j, r)$ are 3×3 *elementary matrices*, and left-multiplication by one of them is an *elementary row operation*. (See **Exercises 11, 15, and 18**.)

Exercises
10. Verify that equation (7) is the result of left-multiplying both sides of equation (6) by the matrix

$$\begin{bmatrix} 1/2 & 0 & 0 \\ 0 & 1 & 0 \\ 0 & 0 & 1 \end{bmatrix}.$$

11. Verify that left-multiplication by $A(i, r)$ of a $3 \times n$ matrix M replaces row i of M with r times row i.

12. Verify that $A(i, r)$ is a nonsingular 3×3 matrix if $r \neq 0$. What is the inverse of $A(i, r)$?

13. How would you define a 4×4 $A(i, r)$ for replacing row i of a $4 \times n$ matrix M with r times row i? An $m \times n$ matrix M?

14. Verify that equation (8) is the result of left-multiplying both sides of equation (7) by the matrix

$$\begin{bmatrix} 1 & 0 & 0 \\ -2 & 1 & 0 \\ 0 & 0 & 1 \end{bmatrix}.$$

15. Verify that left-multiplication by $B(i, j, r)$ of a $3 \times n$ matrix M replaces row i of M with row i plus r times row j ($i \neq j$).

16. Verify that $B(i, j, r)$ is a nonsingular 3×3 matrix if $i \neq j$. What is the inverse of $B(i, j, r)$?

17. How would you define a 4×4 $B(i, j, r)$ for replacing row i of a $4 \times n$ matrix M with row i plus r times row j? An $m \times n$ matrix M?

18. Determine a nonsingular 3×3 matrix $C(i, j)$ such that left-multiplication by $C(i, j)$ of a $3 \times n$ matrix M has the effect of interchanging rows i and j of M. Verify that your $C(i, j)$ is nonsingular; what is the inverse of $C(i, j)$? Extend your result to $m \times n$ matrices M.

    ($C(i, j)$ is the third of the three types of elementary matrix corresponding to an elementary row operation. We will not need it for the simplex method.)

3. The Simplex Method

3.1. Motivation and Discussion

Our goal is to left-multiply both sides of the matrix equation for the nutrition problem (6) by appropriate 3×3 nonsingular elementary matrices, to obtain an equivalent form of (6) that is easier to solve.

I. We will start with some simple values for $x_1, \ldots, x_5$ that satisfy **(6)** and the additional nonnegativity constraints $x_1, \ldots, x_5 \geq 0$. The easiest values that fit all the constraints are the original variables set to zero and the slack variables set to the values on the right-hand side: $x_1 = x_2 = x_3 = 0$, $x_4 = 12$, and $x_5 = 15$. Then c has the value 0. However, c is not at its maximum value, since c would become positive if we let x_1, x_2, or x_3 become positive. Of course, we would then have to reduce x_4 from 12 and x_5 from 15, in order to maintain the constraints $2x_1 + x_2 + 3x_3 + x_4 = 12$ and $2x_1 + 2x_2 + x_3 + x_5 = 15$.

II. In order to increase the value of c, the largest increase per unit is obtained by letting x_1 increase, since x_1 increases c by 10 units per ounce, whereas x_2 and x_3 only increase c by 7 and 8 units per ounce, respectively. With x_2 and x_3 remaining at 0, row 1 of equation **(6)** says $2x_1 + x_4 = 12$, and row 2 says $2x_1 + x_5 = 15$. Since $x_4 \geq 0$ and $x_5 \geq 0$, row 1 requires $x_1 \leq 12/2 = 6$ and row 2 requires $x_1 \leq 15/2$. We want x_1 to satisfy both restrictions, hence we must retain the stronger restriction $x_1 \leq 6$ from row 1. The coefficient we wish to retain is the one in row 1 and column 1 of the left-hand matrix of **(6)**. We will rewrite **(6)** in an equivalent form so column 1 of the left-hand matrix is replaced by the column vector $[1\ 0\ 0]^T$; this operation is called *pivoting* on the element in row 1 and column 1. We accomplish it by left-multiplying both sides of **(6)** by the product of elementary matrices

$$\begin{bmatrix} 1 & 0 & 0 \\ 0 & 1 & 0 \\ -10 & 0 & 1 \end{bmatrix} \begin{bmatrix} 1 & 0 & 0 \\ -2 & 1 & 0 \\ 0 & 0 & 1 \end{bmatrix} \begin{bmatrix} 1/2 & 0 & 0 \\ 0 & 1 & 0 \\ 0 & 0 & 1 \end{bmatrix}.$$

We obtain first **(7)** and then **(8)** of Section 2, and finally

$$\begin{bmatrix} 1 & 1/2 & 3/2 & 1/2 & 0 \\ 0 & 1 & -2 & -1 & 1 \\ 0 & 2 & -7 & -5 & 0 \end{bmatrix} \begin{bmatrix} x_1 \\ x_2 \\ x_3 \\ x_4 \\ x_5 \end{bmatrix} = \begin{bmatrix} 6 \\ 3 \\ c - 60 \end{bmatrix}. \quad (9)$$

Note that first the column and then the row of the pivot value are chosen as follows: The column of the pivot value corresponds to the greatest rate of increase (for a maximization problem), and the row of the pivot value corresponds to the stronger restriction.

III. We now repeat the analysis of steps I and II on equation **(9)**. Now the easiest nonnegative values to fit the constraints are $x_2 = x_3 = x_4 = 0$, $x_1 = 6$, and $x_5 = 3$, giving $c = 60$. From row 3 of equation **(9)**, $c = 60 + 2x_2 - 7x_3 - 5x_4$, hence c can still be increased by letting x_2 increase. The restrictions on x_2 from rows 1 and 2 are $x_2 \leq 6/(1/2) = 12$ and $x_2 \leq 3$, respectively; hence we retain the stronger row 2 restriction on x_2. We now rewrite **(9)** in an equivalent form so column 2 of the left-hand matrix is replaced by the column vector $[0\ 1\ 0]^T$, i.e. we pivot on the element in row 2 and column 2. Therefore we left-multiply both sides of **(9)** by the product of elementary matrices

$$\begin{bmatrix} 1 & 0 & 0 \\ 0 & 1 & 0 \\ 0 & -2 & 1 \end{bmatrix} \begin{bmatrix} 1 & -1/2 & 0 \\ 0 & 1 & 0 \\ 0 & 0 & 1 \end{bmatrix},$$

obtaining

$$\begin{bmatrix} 1 & 0 & 5/2 & 1 & -1/2 \\ 0 & 1 & -2 & -1 & 1 \\ 0 & 0 & -3 & -3 & -2 \end{bmatrix} \begin{bmatrix} x_1 \\ x_2 \\ x_3 \\ x_4 \\ x_5 \end{bmatrix} = \begin{bmatrix} 9/2 \\ 3 \\ c - 66 \end{bmatrix}. \quad (10)$$

IV. Again, we repeat the analysis of steps I and II on equation **(10)**. The easiest nonnegative values to fit the constraints are $x_3 = x_4 = x_5 = 0$, $x_1 = 9/2$, and $x_2 = 3$, giving $c = 66$. From row 3 of **(11)**, $c = 66 - 3x_3 - 3x_4 - 2x_5$; and since x_3, x_4 and x_5 must be ≥ 0, c cannot be increased further. Therefore $c = 66$ is the maximum number of daily protein units possible, achieved with 9/2 ounces of food 1, 3 ounces of food 2, and 0 ounces of food 3.

3.2. Summary

The goal of the simplex method for a maximization problem is to rewrite the original matrix equation in an equivalent form such that the last row of the left-hand matrix contains only zeroes and negative numbers, and such that the last row of the right-hand matrix is c minus a constant. Then the maximum value of c can be readily obtained from the new equivalent matrix equation, as in step IV of Section 3.1. The simplex method for our problem consists of repeating the following two steps until the goal is achieved:

I. Find the simplest (all but two values 0, where two is the number of equality constraints) nonnegative values $x_1, \ldots, x_5$ to satisfy the matrix equation, and determine whether c is already at its maximum.

II. If c is not at its maximum, determine the appropriate pivot element. Choose the largest positive number in the last row of the left-hand matrix, and let j denote its column number. Take the quotients of the elements of the rows of the right-hand matrix except for the last row, over the corresponding row elements in column j of the left-hand matrix; let i denote the row number that gives the smallest positive quotient. Pivot on the element in row i and column j (i.e., replace column j with 0's except for a 1 in row i) of the left-hand matrix via multiplication by appropriate elementary matrices, obtaining a new equivalent matrix equation. (Notice that we never pivot on an element in the last row.)

We have solved a problem of maximizing c. In order to solve a problem of minimizing c, we could first maximize $-c$, and the minimum value of c would then be the negative of the maximum value of $-c$.

Exercises

19. Use the simplex method to solve the linear programming problem of Exercise 7.

20. Use the simplex method to solve the following:

Find the maximum value of
$$c = 5x_1 + 8x_2$$
subject to
$$2x_1 + 5x_2 \leq 40$$
$$x_1 + x_2 \leq 15$$
$$4x_1 + x_2 \leq 48$$
$$x_1, x_2 \geq 0.$$

21. Several interesting applications of linear programming (both examples and exercises), which the reader can now solve, may be found in [Rorres and Anton 1984, Chapters 20, 22].

4. Further Discussion

We have provided a brief look at the simplex method for solving a linear programming problem. However, several important mathematical questions are beyond what we can present here. For example, we have to ask whether repeating steps I and II of Section 3.2 will always eventually lead us to a solution, as well as whether it is always possible to carry out these steps. We should also specify a more rigorous description of them.

In addition, there are variations on the simplex method, such as the revised simplex method and methods related to the dual problem [Luenberger 1973, Noble 1969]. The simplex method also has an interesting geometric interpretation related to convex sets [Luenberger 1973].

5. References

Anton, Howard. 1987. *Elementary Linear Algebra*. 5th ed. New York: Wiley.

> Chapter 1 presents a systematic description of matrix arithmetic, elementary matrices, and Gaussian elimination for solving systems of linear equations.

Dantzig, George B. 1963. *Linear Programming and Extensions*. Princeton, NJ: Princeton University Press.

> Detailed introduction to linear programming and the simplex method, including origins, history, and applications.

Lane, Kenneth D., and John Goulet. 1985. The Karmarkar algorithm: New issues in the computational complexity of linear programming. *The UMAP Journal* 6(3):5–20.

> Provides an excellent summary of a recent exciting development in linear programming, as well as a survey of linear programming in general.

Luenberger, David G. 1973. *Introduction to Linear and Nonlinear Programming*. Reading, MA: Addison-Wesley.

> Chapters 2 and 3 deal with basic properties of linear programming, convexity, the simplex method, and the revised simplex method. Chapter 4 discusses duality and the dual simplex method.

Noble, Ben. 1969. *Applied Linear Algebra*. Englewood Cliffs, NJ: Prentice-Hall.

> Chapter 6 contains an excellent rigorous matrix-based introduction to linear programming and the simplex method, as well as discussions of the role of convexity and the dual problem.

Rorres, Chris, and Howard Anton. 1984. *Applications of Linear Algebra*. 3rd ed. New York: Wiley.

> Chapters 20–22 provide an introduction to linear programming, including geometric interpretation, the simplex method, and applications.

Rosenberg, Nancy S. 1980. *Linear Programming in Two Dimensions: I, II*. UMAP Modules in Undergraduate Mathematics and Its Applications: 453–454. Lexington, MA: COMAP, Inc.

> These UMAP Modules treat linear programming graphically, with the second focusing on sensitivity analysis.

Stigler, George J. 1945. The cost of subsistence. *Journal of Farm Economics* 27:303–314.

Presents the classical "diet problem," minimizing the cost of the diet subject to nutrition constraints.

Strang, Gilbert. 1988. *Linear Algebra and Its Applications.* 3rd ed. New York: Academic.

Chapter 1 includes a presentation of Gaussian elimination and elementary matrices. Linear programming, including the simplex method and duality theory, is presented in Chapter 8.

6. Answers to Exercises

1. Maximum value of c is 19 with $x_1 = 4$, $x_2 = 0$ (see **Figure 2**).

2. Minimum value of c is 3 with $x_1 = 9$, $x_2 = 0$.

3. Maximum value of c is 36 with $x_1 = 0$, $x_2 = 3$.

4. Minimum value of c is 17 with $x_1 = x_2 = 0$.

5. c has no maximum value; c can be arbitrarily large as x_1 and x_2 become arbitrarily large.

6. There are no values for x_1 and x_2 which satisfy these constraints, since $x_1 \geq 0$ and $x_2 \geq 0$ imply $x_1 + x_2 \geq 0$. This problem is called *infeasible*.

7. Let x_1 denote the number of pounds of sugar cookies baked per day, and x_2 the number of pounds of gingerbread cookies. The time constraint is

$$(1/2)x_1 + 1x_2 \leq 12, \quad \text{or} \quad x_1 + 2x_2 \leq 24.$$

The total sales constraint is $x_1 + x_2 \leq 20$. We wish to maximize daily profit, which is $30x_1 + 50x_2$. Hence the problem is:

Find the maximum value of

$$c = 30x_1 + 50x_2$$

subject to

$$x_1 + 2x_2 \leq 24$$
$$x_1 + x_2 \leq 20$$
$$x_1, x_2 \geq 0.$$

(See **Exercise 19** for the simplex method solution.)

8. $3x_1 + x_2 \leq 7$ is equivalent to $3x_1 + x_2 + x_3 = 7$ with $x_3 \geq 0$.

9. $x_1 + 7x_2 \geq 5$ is equivalent to $x_1 + 7x_2 - x_3 = 5$ with $x_3 \geq 0$.

10. $\begin{bmatrix} 1/2 & 0 & 0 \\ 0 & 1 & 0 \\ 0 & 0 & 1 \end{bmatrix} \begin{bmatrix} 2 & 1 & 3 & 1 & 0 \\ 2 & 2 & 1 & 0 & 1 \\ 10 & 7 & 8 & 0 & 0 \end{bmatrix} = \begin{bmatrix} 1 & 1/2 & 3/2 & 1/2 & 0 \\ 2 & 2 & 1 & 0 & 1 \\ 10 & 7 & 8 & 0 & 0 \end{bmatrix},$

$\begin{bmatrix} 1/2 & 0 & 0 \\ 0 & 1 & 0 \\ 0 & 0 & 1 \end{bmatrix} \begin{bmatrix} 12 \\ 15 \\ c \end{bmatrix} = \begin{bmatrix} 6 \\ 15 \\ c \end{bmatrix}.$

11. Since $A(i, r)$ is the identity matrix with the ith diagonal element replaced by r, then all rows except the ith row of M remain the same after left-multiplication by $A(i, r)$, and each element of the ith row of M is multiplied by r.

12. Since $\det A(i, r) = r \neq 0$, then $A(i, r)$ is nonsingular. In fact,

$$A(i, r)^{-1} = A(i, 1/r) \qquad \text{for } r \neq 0.$$

13. For an $m \times n$ matrix M, let $A(i, r)$ be the $m \times m$ identity matrix with the ith diagonal element replaced with r. Then left-multiplication of M by $A(i, r)$ replaces row i of M with r times row i.

14. $\begin{bmatrix} 1 & 0 & 0 \\ -2 & 1 & 0 \\ 0 & 0 & 1 \end{bmatrix} \begin{bmatrix} 1 & 1/2 & 3/2 & 1/2 & 0 \\ 2 & 2 & 1 & 0 & 1 \\ 10 & 7 & 8 & 0 & 0 \end{bmatrix} = \begin{bmatrix} 1 & 1/2 & 3/2 & 1/2 & 0 \\ 0 & 1 & -2 & -1 & 1 \\ 10 & 7 & 8 & 0 & 0 \end{bmatrix},$

$\begin{bmatrix} 1 & 0 & 0 \\ -2 & 1 & 0 \\ 0 & 0 & 1 \end{bmatrix} \begin{bmatrix} 6 \\ 15 \\ c \end{bmatrix} = \begin{bmatrix} 6 \\ 3 \\ c \end{bmatrix}.$

15. Since $B(i, j, r)$ is the identity matrix except in the ith row, then all rows except the ith row of M remain the same after left-multiplication by $B(i, j, r)$. Since $i \neq j$, after left-multiplication by $B(i, j, r)$, row i of M becomes row i plus r times row j, since $B(i, j, r)$ has 1 in row i and column i and the constant r in row i and column j.

16. Det $B(i, j, r) = 1$ if $i \neq j$, hence $B(i, j, r)$ is non-singular for $i \neq j$. In fact, $B(i, j, r)^{-1} = B(i, j, -r)$.

17. For an $m \times n$ matrix M, let $B(i, j, r)$ be the $m \times m$ identity matrix with the element in row i and column j replaced with r $(i \neq j)$. Then left-multiplication of M by $B(i, j, r)$ replaces row i of M with row i plus r times row j.

18. For an $m \times n$ matrix M, let $C(i, j)$ be the $m \times m$ identity matrix with rows i and j interchanged. Then left-multiplication of M by $C(i, j)$ interchanges rows i and j of M. Det $C(i, j) = -1$ for $i \neq j$, and $C(i, j)^{-1} = C(j, i)$.

19. Maximize c subject to

$$\begin{bmatrix} 1 & 2 & 1 & 0 \\ 1 & 1 & 0 & 1 \\ 30 & 50 & 0 & 0 \end{bmatrix} \begin{bmatrix} x_1 \\ x_2 \\ x_3 \\ x_4 \end{bmatrix} = \begin{bmatrix} 24 \\ 20 \\ c \end{bmatrix}$$

and $x_1, x_2, x_3, x_4 \geq 0$ (x_3 and x_4 are slack variables).

Linear Programming Via Elementary Matrices   311

I. Start with $x_1 = x_2 = 0$, $c = 0$, $x_3 = 24$, $x_4 = 20$.
II. Pivot on row 1 column 2, obtaining

$$\begin{bmatrix} 1/2 & 1 & 1/2 & 0 \\ 1/2 & 0 & -1/2 & 1 \\ 5 & 0 & -25 & 0 \end{bmatrix} \begin{bmatrix} x_1 \\ x_2 \\ x_3 \\ x_4 \end{bmatrix} = \begin{bmatrix} 12 \\ 8 \\ c - 600 \end{bmatrix}.$$

I. New values: $x_1 = x_3 = 0$, $c = 600$, $x_2 = 12$, $x_4 = 8$.
II. Pivot on row 2 column 1, obtaining

$$\begin{bmatrix} 0 & 1 & 1 & -1 \\ 1 & 0 & -1 & 2 \\ 0 & 0 & -20 & -10 \end{bmatrix} \begin{bmatrix} x_1 \\ x_2 \\ x_3 \\ x_4 \end{bmatrix} = \begin{bmatrix} 4 \\ 16 \\ c - 680 \end{bmatrix}.$$

I. New values: $x_3 = x_4 = 0$, $c = 680$, $x_1 = 16$, $x_2 = 4$. Since

$$c = 680 - 20x_3 - 10x_4 \quad \text{and} \quad x_3, x_4 \geq 0,$$

then $c = 680$ is maximum.
Therefore, the maximum profit per day is $6.80, with 16 pounds of sugar cookies, and 4 pounds of gingerbread cookies.

20. Maximize c subject to

$$\begin{bmatrix} 2 & 5 & 1 & 0 & 0 \\ 1 & 1 & 0 & 1 & 0 \\ 4 & 1 & 0 & 0 & 1 \\ 5 & 8 & 0 & 0 & 0 \end{bmatrix} \begin{bmatrix} x_1 \\ x_2 \\ x_3 \\ x_4 \\ x_5 \end{bmatrix} = \begin{bmatrix} 40 \\ 15 \\ 48 \\ c \end{bmatrix}$$

and $x_1, \ldots, x_5 \geq 0$ (x_3, x_4 and x_5 are slack variables).

I. Start with $x_1 = x_2 = 0$, $c = 0$, $x_3 = 40$, $x_4 = 15$, $x_5 = 48$.
II. Pivot on row 1 column 2, obtaining

$$\begin{bmatrix} 2/5 & 1 & 1/5 & 0 & 0 \\ 3/5 & 0 & -1/5 & 1 & 0 \\ 18/5 & 0 & -1/5 & 0 & 1 \\ .9/5 & 0 & -8/5 & 0 & 0 \end{bmatrix} \begin{bmatrix} x_1 \\ x_2 \\ x_3 \\ x_4 \\ x_5 \end{bmatrix} = \begin{bmatrix} 8 \\ 7 \\ 40 \\ c - 64 \end{bmatrix}.$$

I. New values: $x_1 = x_3 = 0$, $c = 64$, $x_2 = 8$, $x_4 = 7$, $x_5 = 40$.
II. Pivot on row 3 column 1, obtaining

$$\begin{bmatrix} 0 & 1 & 2/9 & 0 & -1/9 \\ 0 & 0 & -1/6 & 1 & -1/6 \\ 1 & 0 & -1/18 & 0 & 5/18 \\ 0 & 0 & -3/2 & 0 & -1/2 \end{bmatrix} \begin{bmatrix} x_1 \\ x_2 \\ x_3 \\ x_4 \\ x_5 \end{bmatrix} = \begin{bmatrix} 32/9 \\ 1/3 \\ 100/9 \\ c - 84 \end{bmatrix}.$$

I. New values: $x_3 = x_5 = 0$, $c = 84$, $x_1 = 100/9$, $x_2 = 32/9$, $x_4 = 1/3$. Since $c = 84 - 3/2x_3 - (1/2)x_5$ and $x_3, x_5 \geq 0$, then $c = 84$ is maximum.
Therefore the maximum value of c is 84, with $x_1 = 100/9$ and $x_2 = 32/9$. There is positive slack of $1/3$ in the second constraint $x_1 + x_2 \leq 15$, and zero slack in the first and third constraints.

QA 37.2 .U45 1987

UMAP modules

APR 2 4 1989